HIGH
STATIC
DEAD
LINES

High Static, Dead Lines: Sonic Spectres and the Object Hereafter
by Kristen Gallerneaux
Published by Strange Attractor Press 2018. Second printing, 2019.
ISBN: 9781907222665

Cover image and illustrations by Kristen Gallerneaux
Design by Jamie Sutcliffe

Kristen Gallerneaux has asserted her moral right to be identified as the author of this work in accordance with the Copyright, Designs and Patents Act, 1988. All rights reserved. No part of this publication may be reproduced in any form or by any means without the written permission of the publishers. A CIP catalogue record for this book is available from the British Library.

Strange Attractor Press
BM SAP, London, WC1N 3XX, UK
www.strangeattractor.co.uk

Distributed by The MIT Press, Cambridge, Massachusetts.
And London, England.
Printed and bound in the UK by TJ International, Padstow.

HIGH STATIC

Sonic Spectres and the Object Hereafter

DEAD LINES

KRISTEN GALLERNEAUX

TABLE OF CONTENTS

Foreword: Don't Go Chasing Heads Over Waterfalls
by Dave Tompkins..*ix*

Introduction..1

I. DEAD LINES

Electric Line Events..19

We Are Not Yet Experts: The Mimetic Dead,
Paranormal Acoustics, And The Medium Jesse Shepard.........25

Elizabeth Street Apports..53

II. UP THERE

Nothing Is Static In The Tides:
From The Sea To Mars With Charles Francis Jenkins.................57

Book Street Skies..75

Squier's Music For The Trees......................................79

Radio Cadiz: Up the Hatch With The Piccards.........................85

III. FREQUENCIES

Tune In To The Tummy: Vladimir Zworykin's Radio Pill.......93

Charles Apgar & Sordid Frequencies..97

Calling, Haunting..103

All Together Now, Speaking On A Beam of Light:
An Impure History Of The Maser And Laser..........................105

Hello My Name Is Votrax..127

Fetch X (Living Ghost Lifts Up The Receiver)..........................139

Rush In Reverse...151

IV. BROADCAST

We Haunt You While You Watch: TV Hijackings,
The Eerie In The Middle, And Cochlear Control...................155

War, Wire, Magnets...179

We Bridge The Dead / Breakthrough..185

V. PLAYING THE SPECTRE

Humming Through Salt:
AUDINT's Spurious Frequencies..193

Prison Radio / Shotgun...211

Performance Rituals: The Moog Prototype.............................217

Dirt Synth: Excavating The Sounds Of The Poltergeist..........227

VI. ANCHORS

Prospector Stake From A Mythical Lake.................................233

The Domes: Aural Fieldwork At The End Of The Line........241

BlackBridge...245

Throwing The Apparition Of A Brick Through A Real Ghost:
A Conversation With The Author *by Dave Tompkins*...............257

Bibliography..267
Index...285
Acknowledgements..300

FOREWORD: DON'T GO CHASING HEADS OVER WATERFALLS

by Dave Tompkins

A coffin handle dawdles in the cattail mats of the Sydenham River. Headed southwest for Detroit, it's stuck outside of Ontario, despite the current's intentions. Goby and perch are drawn to the brass. Leeches can't get traction. A crayfish vanishes in a puff of mud.

A curious tug, followed by a snap vibration on the nylon line. Jerked out of the bulrush, the handle is dragged across the water through fractured light. The girl reels it into her grandfather's rowboat. She wants to keep it, less for its shiny weight than for a misplaced attachment—a reminder of a chest of drawers she lost to a fire. The charred purple Count doll. The "Baby Bat" kite with the screaming eyeballs that once dove through the sky bloodshot, looking more hungover than scary.

Reeds whistle and clatter. Her grandfather looks at the handle, wipes off the algae. No stranger to worms, polished to be buried, only seen in the memory of his pallbearer grip, his hands scarred from working an assembly line for a manufacturer of Happy Meal glasses. Sometimes he'd bring home the defectives. The Hamburglar, just slightly off.

He makes her toss the handle back into the silt, as if the rest of the coffin had gotten away. Actually, several had.

Earlier that week, the Riverside Cemetery flooded. The barrier wall separating the dead from the Sydenham riverbank had burst, sending a fleet of coffins bobbing downstream. Kristen Gallerneaux had been to that cemetery before, a goth with a Nikon trying to capture headstones. Her mother watched from the powder-blue Plymouth hatchback, leaning on the horn when she saw a wraith shadowing her teenager. Their family of Canadian spiritualists kept markers at Riverview Cemetery but had their ashes interred, skipping the regatta of the dead. According to Gallerneaux family lore, spirits don't really hang graveside, so why splurge on a vessel? Coffins are for comforting the undead anyway.

Kristen once described an usher she met in a dream as "very hearse driver from *Burnt Offerings*." In *High Static, Dead Lines*, she writes of emergency phones that were once installed in coffins in case of premature burial—folklore and legend seeded by technology. Invention can be driven by mourning, beholden to imagination, if even (and often) in failure. Belief is proof of the human spirit more than the supernatural. After all, somebody had to come up with this crazy shit. Those engineered radio pills, "howl arresters," and lasers patented for exercising your cat, if not driving it mad. You could get a call from the future, your deceased helicopter-mechanic aunt on the line. Belief is in the dirt, the grounded materiality. Salt, stone, crystal, soil. A glint of mineral transduction.

It can be found in the bone-derived collagen used in film stock, buried in a gelatinous footprint belonging to *The Incredible Melting Man* (1977). We find ourselves on another riverbank, next to an astronaut in a hospital robe. He suffers from a rare skin deliquescence acquired from Saturn, a condition so upsetting that he decapitates a fisherman. He tosses the head in the creek like a turned cabbage, then steps on the fisherman's sandwich and storms off into the brush. Well then! The camera sticks around by the water to follow the journey of the fisherman's head downstream while the Melting Man is up to lord knows what. May as well, it's nice out. This lollygag of screen-time is in full appreciation of the day, its gurgle and birdsong, all the spring finery. The head takes an idyllic bob through eddies of leaves, catching twigs and skiffs of bark, scattering gator ticks and, in a grand flourish, is sent over a waterfall. It all could've (and probably should've) ended there. But the head sails on in the B-roll of our memory, probably annoyed at this point, its expression more Garfield than Horrified.

This scene continues in Gallerneaux's admiration, in the wonder that it actually exists, conveyed to me from the front seat as her husband Bernie drives us out to a Red Lobster in Ypsilanti, Michigan. The voyage of the fisherman's head is an agent of calm, returning her to the forsaken open spaces of her childhood in rural Ontario. It's a story that could only be told by this lone eBay-bidder of *The Incredible Melting Man* script, purchased in reverence of a screen cue so peculiar it draws attention to its half-bitten typeface: an incredible melting foot stepping on a PB&J. A squish for a squish. It appears in the mail with a note from director William Sachs. "This is the only one." As if ever.

I loiter on the fisherman's lunch. Should the five-second rule be amended for a sandwich bearing a gooey footprint from Saturn? But there's no time. The conversation has leapt to psychic photography. Gallerneaux asks if I'm familiar with the strange case of Ted Serios, the alcoholic bellhop from Chicago who could project the image of an Olmec stela from his mind onto Polaroid film. The print remains in the collection of a psychiatrist named Wormington. Gallerneaux supposes Dr. Wormington could be the third-party owner of a brain-blast photo. Her favorite image of Serios's "Thoughtography" is one with the word "Canadian" misspelled on a hangar used by the air division of the Royal Canadian Mounted Police. She reasons that Serios was semi-illiterate, so it's possible his mind-camera scrambled the letters. "Who would go to the trouble of faking that?" she asks. We turn into the Red Lobster parking lot, past an exposed sewer pipe under construction. Gallerneaux mentions that her friend's mom had a fatal cameo in the cannibal effluent film *C.H.U.D.* (1984).

We're at the seafood chain to meet Rick Davis, co-founder of Cybotron and creator of electrofunk in Detroit in the early 80s. I'll always be head-over-waterfalls for tracks like "Clear" and "Alleys of Your Mind." (Detroit rapper Danny Brown once considered calling his cat Cybotron: "That'd have been a tight kitty name.") There's very little thoughtographic evidence of Davis online, so we enter Red Lobster looking for the black Vietnam vet most likely to own *The Incredible Melting Man* on VHS. He's hunched over the bar, a giant in a black leather motorcycle jacket and camo pants, drinking a Bahama Mama. The rest of the room, the bartender and customer ambience, seems staged and out of scale.

I hand Davis the *Wicker Man* issue of *Famous Monsters of Filmland* and a still from the Ray Harryhausen film *Golden Voyage of Sinbad*. Some burnt offerings. He smiles behind wraparound sunglasses. The centaur and the gryphon appear to be bickering. Are you sure? We order cheese biscuits. Gallerneaux reveals that she was recently the sore loser at an auction for a collection of Harryhausen's Super 8 cameras. Davis says he ranks the Melting Man decapitation as one of the best of its time. He then uses his cloth napkin to do an impression of Caltiki the Immortal Monster, once described by the film critic Phil Hardy as "a huge quantity of cow entrails with a man inside." Davis thinks there were molasses involved. It's Saturday night in Ypsilanti and we're all clearly rooting for the monster. The fisherman's head could've been a tallboy.

That night I slept in Gallerneaux's library in Dearborn. On one shelf is a projectile of drywall that attacked her husband Bernie while they moved into their previous residence in Detroit. She refers to it as "an imprint of defeat." Bernie won't allow certain things in the house. *The Gingerdead Man*, starring Gary Busey, for instance. But the plaster remains. Also on the bookshelf are *The World of Ted Serios*; Roger de Lafforest's *Houses That Kill*; Denis Gifford's *A Pictorial History of Horror Movies*; a book by the Latvian parapsychologist Raudive; journals of psychical research; old issues of *Famous Monsters* and *OMNI*; comics stolen from older brothers; *Wisconsin Death Trip*; Allen Shelton's *Dreamworlds of Alabama*; and a couple of dirt-synth dime bags, sourced from crisis sites of apportation.

Reading *High Static, Dead Lines*, I recognize stories filtered through brief flashes of memory, the glitter of belly-up scales on the Sydenham. The head going over the waterfall becomes a zombie tumbling from an escalator into a fountain at a mall in Pittsburgh. (Gallerneaux used this scene from *Dawn of the Dead* for a workshop titled "Making Good Beats out of Bad Muzak.") Or a bass drum flying from the window of a castle tower in Cortachy, Scotland in the 1700s, with the castle's resident drummer-boy sealed inside the membrane. Did the plummet allow for one final kick? Did Gallerneaux really make a beat from a Kmart intercom in Gulfport, Mississippi on the day the store closed? (There was a special on reverb that day.) Other stories concern abandoned technologies that have been deposited at her desk at The Henry Ford in Dearborn, where daydreams are left to obsolescence in the basement, while children marvel at the gleam of Model 500s and toy trains displayed on the floors above.

The winding hallway to the Henry Ford vaults has been known to trap smells from the employee dining room. Today it's cabbage soup. Fluorescent tubes hum to our foreheads as we take in the sulfides. The walls and floors are clinically white. I almost expect to see the Incredible Melting Man turn the corner, dripping red globs in his terry cloth hospital robe. Instead it's a woman in a Civil War bonnet. She glides toward us with a wicker basket under her arm, her blue plaid dress floating on a wire hoop cage. On "Civil War Days," woolen Confederates often stray from the battlefield to wander the museum for air-conditioning and peer into Kennedy's last limo. The woman in the bonnet was looking for a 21st century bathroom. She floats on down the hall *en route* to Owl Creek Bridge, another prime fishing spot. Gallerneaux leads us into the archives, a system of rolling shelves opened by lever. She inserts a safety pin into the handle, a precaution to keep us from getting crushed by walls containing Happy Meal toys and the Abe Lincoln death mask. There's much to see here. A mechanical talking lion, a Mr. Bill T-shirt, a magic lantern depicting a waterfall from Hell, Michigan.

My favorite is the prison monkey casino (est. 1914), which sits inside a 16-by-20 glass case, next to the Moog synthesizer prototype. Carved and glued in the Massachusetts state prison in Charlestown, the monkey bar was given to Henry Ford anonymously, "a gift from a prisoner." The monkeys themselves have been knifed out of peach pits and scraps of wood, detailed to the elbow. They play cards, drink and discuss what's next, what we'll never see. Two monkeys lounge on the floor behind the piano, smoking opium from tiny pipes. Across the room, a police monkey takes a bribe. They might want a ticket to the giant Moog spaceship in the diamond mirror above the bar.

We squint into the murmur. What could they be talking about? Failed conversations with Mars via a WWII field radio in Aisle C300. Huckster séances used for crooked real-estate schemes in San Diego. Tragic residue? The monkeys who speak no evil could just type it out on the Votrax, a tactile phoneme device that sits mute in the Henry Ford archives, waiting to be activated by the cuss of Q*bert in the pages ahead. The casino seems to be in stop motion, waiting to resume its seedy business, a future framed in a thousand tiny movements. The piano player does not turn around. He watches the mirror, waiting for our reflections to disappear.

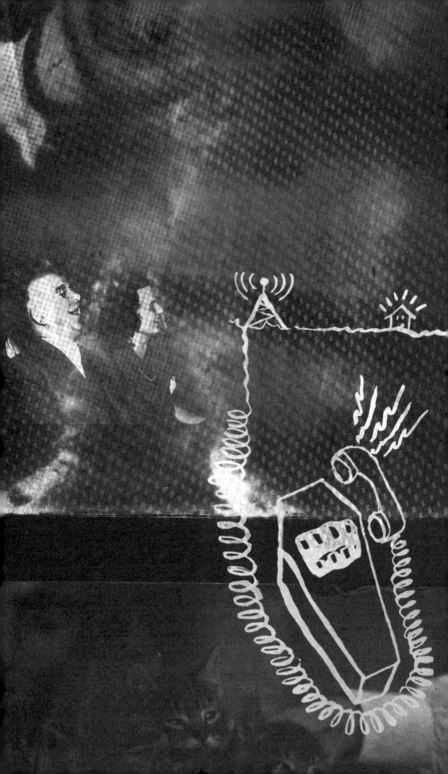

INTRODUCTION

I. *Summoning the Sonic Spectre*

When I was eight years old, my mother and I lived in a house in rural Ontario that was rumored to be haunted. Two weeks after we moved in, I developed a double-dose case of chicken pox and the measles. During my month-long quarantine, when my mother left to start her night shift as a waitress and my uncle had not yet come home from the brass factory, I was often left alone in the house. There simply wasn't enough money for a babysitter, and they likely thought I would sleep through the changeover, which I rarely did. The calming din of WOMC-FM, Detroit's "Oldies 104.3," was always left on in the living room as surrogate comfort in case I woke up. Underneath the warm radio static, it was the indisputable sound of silverware rattling away in the kitchen drawers that was the usual culprit for my awakening. Other nights, the phone rang for what seemed like hours, but whenever I picked up the receiver, the line was all dead air.

Late one night, bored from weeks of forced bed rest, I decided to brave a walkabout through the house. "I'll Turn To Stone" by the Four Tops played on the radio, anticipating the events that would soon follow. Knowing I was alone in the house, I was able to act on a compulsion to revisit the contents stored in the underbelly of an old wooden bench in the foyer. It's important to note something here: When we moved into the house, we essentially moved into a mausoleum. The house belonged to my mother's new boyfriend, a schoolteacher who worked in a remote mill town nine hours north of us. His parents had died in quick succession, and he'd never made the effort to deal with the estate sale. So the house on Elizabeth Street was full of the stuff of everyday life: furniture, kitchenware, a piano, books. There was moldering food in the refrigerator, expired medication in the bathroom cabinets, and an ancient load of laundry left in the washing machine, which had compressed into a uniform cylinder of rotten, stinking fibers. Whenever we used the machine to wash our clothes, we contaminated ourselves with the sharp scent of mold, melding just a little bit more with the stubborn agency of the house. For at least five years the house had been shuttered. It was unlived-in for a decade, until my mother's boyfriend handed us the keys and we crossed its threshold.

As I rifled through the bench's contents that night, I did so under the sneaking spirit of discovery that colored life in that house, living alongside the *unheimlich* traces of someone else's life, always feeling like a guest. When I first flipped open the bench's hinged seat, I found stacks of yellowing papers and Ditto-copied sheet music. The real object of my obsession, however, was a weird flexi disc stuffed between the pages of a self-help magazine from the 1960s. The record was thin—as flexible as the magazine itself. I was compelled to rip the record out along its perforated edge and carry it to the toy turntable in my bedroom.

As a toddler, I was infatuated with the country records my uncle played, and I would try to make his vinyl resonate by spinning the LPs on one finger, using my other finger as though it were a stylus while singing made-up lyrics to "Your Cheatin' Heart" as the adults laughed. By now, I was old enough to know I needed a record player to release what was hidden in the grooves. I needed to hear what was on that record, and I finally had the privacy to do so.

Finding a music cabinet in the parlor, I put the thin piece of vinyl on the turntable and dropped the needle. The scratch and pop that followed shifted my state of mind to prepare for whatever the record contained, but I was not equipped to process what happened next. The sounds bleeding from the speakers sounded demonic, like the beginning of Black Sabbath's "Iron Man," which my older brothers (then living with my father) would play at full volume to terrorize me while they flicked my bedroom lights off and on for an extra-distressing strobe effect. In retrospect, this is hilarious. But being left to my own devices in an apparently haunted house was preferable to the extreme teasing I suffered under the rule of my hesher brothers. The slow, chanting voice of the man on the flexi disc was slurring out incantations and commanding me to do things in slow motion: "Lissstennn nnnowww... Rrrreeelaxxx yourrrr thhhhoughtsssss...." Stuck in place for an eternity, I suddenly regained the necessary footing to sprint toward the bedroom and dive under the sheets, putting maximum distance between myself and the sounds coming from the record cabinet.

The record in the magazine was, of course, a 45RPM disc—but the turntable was set to $33^{1/3}$. A simple enough technical fix, but it seemed as though a passé archetype born out of a horror film was unleashed, the recording giving it a space to become more concrete. Set to the right speed—under those circumstances, in that house—the exchange of sound in space might have felt the same. I'd learned a lesson early. Playing a record backwards, letting a platter of lock-groove vinyl loop, testing out the variability of songs playable at more than one speed, the perfectly timed skip—these things all "thicken" sonic media. They can create the feeling of sound coagulating into something protoplasmic, primitive sites for imagined conjurings to manifest with a greater efficacy. Several stories within this book are rooted to the Elizabeth Street house in rural Ontario, but this one is necessary to lay the groundwork for the idea of the "sonic spectre."

Finding ways to allow our media to haunt us is crucial to understanding it. Embracing these sorts of "thick moments" bring about subjective emotional responses. Reconnect the disconnect. Historically, sound waves are the medium that joins ethereal and real dimensions, acting as unnerving translators capable of carrying messages between the living and the dead. "Media always already provide the appearances of spectres," according to author and media theorist Friedrich Kittler.[1] An offhand remark made by Thomas Edison went too far: An alternate use for his phonograph might be to record the last words of the dying, using cylindrical wax tubes to capture the voice that precedes the death rattle. Electrical scientist and physician William Watson plucked fur from his beloved taxidermied pet cat to use in his electrical friction machine. Cat static. Emergency telephones, buried in coffins. Phobic lifelines for premature burials. We listen to the laugh tracks of people who are long gone. This is the habitat of the sonic spectre.

The manifestations of the sonic spectre are interwoven with the histories of media and material culture. They may take the form of innocuous wooden sticks planted near a lake, revived a century later through sonification. Or an entirely different stick, pounded upon the earth, which prompted a moment of sonic terror and became a defining experience for the co-inventor of the Moog synthesizer. The sonic spectre may latch itself to minuscule and obscure pieces of material history—things that may seem unknowable and mysterious, things that become reinstated into the present through

---

[1] Kittler, Friedrich A. *Gramophone, Film, Typewriter* (Stanford, Calif: Stanford University Press, 1999), 12.

modern speculation, discovery, and description. They may form narratives around commercially available acoustic devices, ones "made strange" through supernatural interaction. Architecture may swallow its own insidious sound history, hiding it in plain sight. We might reactivate it by yelling an infinite echo against the ceiling of an abandoned computer factory, or walk through a room that once hosted musical séances. Or we might choose to turn ourselves into broadcast ghosts by swallowing a radio pill or allowing our voices to rocket through the air on beams of MASER light. We might terrorize a city by hijacking the airwaves of a Chicago television station.

The entries found within this book are not intended to be generalities of "the haunted," and they do not dive deeply into the critical densities of paranormal culture and the nature of belief. These occurrences have been written about enough over the last decade by myself and others. There's also a danger in forcing any and all "serious writing" about spectralities to solely act out within the confines of the polished epistemologies of hauntology, psychoanalytic theory, or theories of capitalism. Or, on the other end of the spectrum, to reduce the power of the spectre (symbolic or actual) by filing its presence under the broad category of "the metaphysical." Of course, all of the vetoes listed above will appear throughout this book as contradictions, but will not overtake or posit that any one ontology is superior to any other. This book leans more toward the poetics of wild folklore than rational academese, alternating between the structures of personal narrative, historical encounters, and the essay. Sometimes, the different modes blur together as fictocriticism. The essays, object studies, autobiographical entries, and visual expressions are arranged here to evoke a loose composition—a network of ley lines for the sonic spectre to travel along, and in between.

The first section, *Dead Lines*, might be thought of as the "OG" approach to hauntings: country séances, city séances, and ghost cakes. The second section, *Up There*, rises to skyborn segues with extraterrestrial radios, tree antennas, and stratospheric explorations. Next, in *Frequencies*, sound waves are swallowed, hacked, beamed, crunched through computers, and used as quasi-fictional lifelines. In *Broadcast*, the ghosts of pure media manifest through televised interruptions and a textual, free-form séance. Section five, *Playing the Spectre*, examines the collision between creativity, performance, and the reach of the sonic spectre—from the exhibitions of the British collective AUDINT, to synthesized music and its connection to hauntology. And lastly, in *Anchors*, the sonic spectre is linked to the natural and built environment. It's in the lies of wooden prospector's stakes, and within the perpetual echoes in a dark and storied Canadian forest, or in buildings evading death in the Sonoran desert.

II. *Draining Away Opacities*

Sometime in the autumn of 2013, just a few months after starting my job as a museum curator of technology collections at the Henry Ford Museum, I received a garbled phone message. Someone was mumbling something about "got your number from this woman" followed by a guttural chain of sounds in which I was able to pick out "want to donate this cell phone." I called the number back, and DJ "Uncle" Russ Gibb answered. For years, I'd hoped to discover the whereabouts of the microphone from WKNR-FM—the same equipment that filtered the spoken rumor of Paul McCartney's supposed death. But he didn't know. He did, however, have a 1970s-era suitcase-sized mobile phone that his former roommate, Eric Clapton, once threw in the Detroit River while in a rage with him. Gibb fished it out before it bobbed away, and he said it worked "just fine" once he unscrewed the earpiece and allowed the river to drain out.

The objects under my curatorial care are essentially a huge collection of Latourian black boxes. They exist as physical proof that the more seamless and successful a technology is, the more mystifying and opaque its inner functions become to the everyday user. Objects may develop lives and stories of their own, but they are self-obfuscating. With unrestricted access to an incredible archive of technology that lives just down the hall from my office, I endeavor to reveal resonance within the collections, wringing out forensic-level details in order to broaden understanding and expose impact beyond the allure of sleek shells (or messy tubes and wires). I have immersed myself in the minutiae of the sometimes foreign-seeming language of communications technology: variable condensers, polar relays, wavemeters, howl arresters, superheterodyne transceivers, and galena crystals. The challenge of studying the physical history of media, information, and communication is in knowing how to draw its scattered data back together again, and how to weave a story out of it, to make it accessible—all the while rooting it back to the object in question. Curators collect to neutralize the past, but we also collect the future in the present.

Sonic spectres run rampant through the history of sound, and sonic artifacts haunt daily life at the museum. Out of all of the collections under my stewardship—ranging from histories of computing, television, radio, film, photography, printmaking, and graphic communication—I have had some of the most palpable moments with sound reproduction artifacts. As is evident in the autobiographical recollections that interweave themselves throughout this book, this is not surprising, considering that sound has often served as the catalyst that locks a memory into place. The objects discussed in the chapters that follow, some

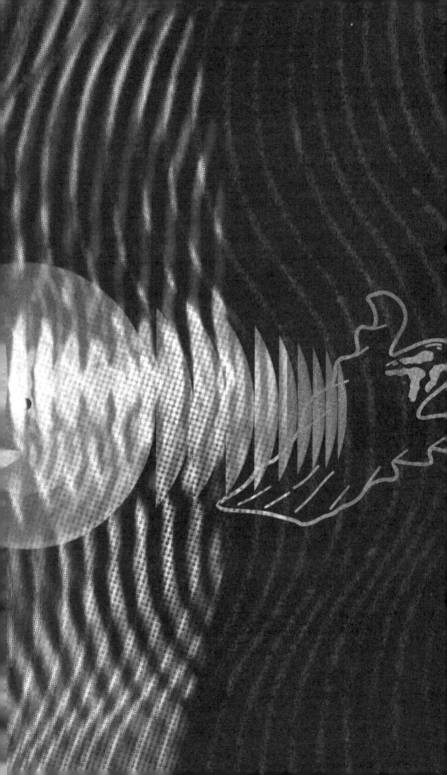

of which may have been silenced for over a century, vibrate with spooky resonance. Through their transition into becoming museum artifacts, they have become *more* alive. In their silence, they seem to hold back secrets. *They are*, and yet, *they aren't*. They are present, and yet they shimmer at the edge of vision like ghosts. Black noise, waiting for someone to pay attention.

III. *Whistling Into Walls*

The writings of David E.H. Jones, produced within the subfield of sound studies known as "archaeoacoustics," are often met with a raised eyebrow. While the study of ancient musical systems found at archaeological sites is considered legitimate, the idea of teasing historic sound vibrations imprinted on the surface of ancient clay pots, or the plaster trowel wielded by a whistling laborer embossing his song into the walls of a building—these are ideas that are too eccentric for most to take seriously.[2] Which isn't to say that *some* kind of sound, *some* evidence of the conditions of time and place upon an object couldn't be recovered.[3]

The creation of the "stone tape theory" is often linked to a book titled *Ghost and Ghoul* (1961) by Thomas Charles Lethbridge, an archaeologist who later abandoned his field to become a full-time parapsychologist. His theory posits that stone architecture and natural materials such as quartz have the ability to serve as recording devices, capturing historical traumas that can later be teased out by contemporary bodies, who act as organic

[2] Jones, David E.H. *The Inventions of Daedalus: A Compendium of Plausible Schemes* (W.H. Freeman & Company), 1982.

[3] See Feaster, Patrick. *Pictures of Sound: One Thousand Years of Educed Audio: 980-1980* (Dust to Digital, 2012).

detectors, decoders, and amplifiers.[4] Architecture may "give up its ghosts," if only momentarily, and always randomly. For Lethbridge, ghosts are residual recordings of the environment, spectral visions on an idiot playback loop, unaware of their own demise. This theory gained attention when the BBC aired Nigel Kneale's *The Stone Tape* on prime-time television in 1972. In this teleplay, characters attempt to force a conversation out from the walls of a haunted castle via a powerful technological intervention. The communication proves to be one-sided, as a complex computer setup, flashing lights, and shrill sonic blast help guide the screams of a dying woman out of the surface of the stone and into the room.

The same cacophony of residual voices was addressed in the 19th century by Sir William Barrett, a member of the Society for Psychical Research:

> *In certain cases of hauntings and apparitions, some kind of local imprint, on material structures or places, has been left by some past events occurring to certain persons, who when on Earth, lived or were closely connected with that particular locality; an echo or phantom of these events becoming perceptible to those now living.*[5]

In 1959, the Swedish artist Friedrich Jürgenson popularized the modern concept of "electronic voice phenomena" (EVP). The standard story goes as follows: After recording birdsong, upon playback, Jürgenson could hear the voice of his dead wife calling out to him—a singular wavelength within the chirpy mix. His paranormal acoustic experiments were some of the first to make use of a commercially available reel-to-reel recorder,

---

4 Lethbridge, Thomas C. *Ghost and Ghoul* (London: Routledge & Kegan Paul, 1961), 75.

5 Barrett, William F. *Psychical Research* (London, Williams & Norgate: 1911), 197.

essentially transforming it into an electronic channeling device. Miles upon miles of magnetic tape rolled gracefully by with the hope of recovering the fragmented conversations of the dead. Using Jürgenson's methods, Latvian parapsychologist Konstantin Raudive collected over 100,000 recordings, portions of which were transcribed and published as *Breakthrough: An Amazing Experiment in Electronic Communication with the Dead.* All of that time, the sonic spectre was just waiting for recording media to improve enough so that it could be heard—dead chatter running in constant parallel with the sounds of everyday life.

We can stare a little deeper into the granite. While these historical psychical research pioneers and rogue archaeologists suggest that rocks can record memory, among common superstitions, there are patterns of storytelling and rituals of banishment that position mineral elements as holding conversational power, too. Driving out spirits can be done by way of the kitchen salt shaker, granular protection cast around a house in an unbroken barricade. Scraping brick walls in a cellar to harvest red dust, laid in lines over thresholds. Limestone along with quartz and magnetite have been loosely cited as geological spirit sinks in residual hauntings long before this idea was diluted by the trend in televised ghost hunting. In the 1974 Italian science-fiction film, *Morel's Invention*, the shipwrecked protagonist spends the first half of the movie stumbling through a wasteland of modernism, dividing his time between a dust-covered museum built in the Brutalist style with an ominous engine room in the basement, and a desolate outdoor landscape. His clothing is the same tonal range as the scrubby desert through which he shuffles, unnecessary camouflage against the "new people" who suddenly appear. He has a one-way mirror view. He can see them, but they can't see him. He is a residual ghost, scratching out his time in a mirrored reality by making hash marks with a rock on a cave wall. Strange vibrations ensue.

IV. *Gut Feelings / Troubled Things*

When objects congregate, either willfully or forcefully (as they are given space to do within this book), they produce theories of the *collective thing*. John Law[6] and Jani Scandura[7] both pose interesting relationships with the animate, archival, and abject qualities of "the mess," as does Raiford Guins. In his book *Game After*, Guins gives new meaning to the rubbish heaps of video game disposal sites.[8] Gay Hawkins puts Bill Brown's theories to use in her essay "Sad Chairs." She claims that when objects are non-compliant, they "reveal their 'thingness,' and they shock us into an awareness…"[9] In *Paraphernalia: The Curious Lives of Magical Things,* Steven Connor's entangled writing style embodies "the mess" as he situates the power of lively objects and their embedded "affordances" (as per James J. Gibson) into their metaphorical networks: "The pipe is a magical compromise between visibility and invisibility. Conspicuously, sometimes obscenely visible though they are themselves, pipes nevertheless conduct a secret ministry. … The network of pipes, like the network of wires, turns a landscape, with all its jagged accident and lumpy irregularity, into an idea."[10] Appropriately, material culture is the site where the ethereal nature of the supernatural and the oscillating effect of philosophical thought unite.

---

6 See Law, John. *After Method: Mess in Social Science Research* (London: Routledge, 2004).

7 See Scandura, Jani. *Down in the Dumps: Place, Modernity, American Depression* (Durham: Duke University Press, 2008).

8 See Guins, Raiford. *Game After: A Cultural Study of Video Game Afterlife* (MIT Press, 2014).

9 Guins 227.

10 Connor, Steven. *Paraphernalia: The Curious Lives of Magical Things* (London: Profile Books, 2011), 151–52.

Christopher Tilley's phenomenological study of the materiality of stone[11] and Jeffrey Cohen's findings in *Stone: An Ecology of the Inhuman* can be extended into otherworldly realms to address subjects like the aforementioned stone tape theory, as well as residual hauntings, or the phenomena of *lithobolia*—those stones, coal, or mud clods that rain onto the roofs of houses afflicted by the poltergeist.

The overlap between critical object studies and paranormal events most often occurs through the discussion of technological development in the mid-19th and early 20th centuries. When the public adopted new forms of communication such as the telegraph, phonograph, telephone, or photograph, the collisions between belief, reproduction, and creativity became permanently altered. While the telegraph and radio *closed* distances on a person-to-person level, they also underwent a simultaneous period of speculative application, to attempt to creatively *extend* human capabilities. Just as sound recording devices are used by investigators to attempt a connection with the voices of the dead, an everyday camera can circumvent its intended function when the images of "the departed" begin to show up on negatives. As technology has become entangled with faith, its various devices have served dual functions—as lifelines to tap into the perceived beyond.

In *Gramophone, Film, Typewriter*, Friedrich Kittler portrays the alternative lives of technologies as a place where "the dead and ghosts become technologically reproducible."[12] In *Dumbstruck*, Steven Connor's phenomenological study of voice and disembodiment, he notes that the séance lets us hear "the workings of the *machine in the ghost*" (emphasis mine) and that

---

11 See Tilley, Christopher and Wayne Bennett. *The Materiality of Stone: Vol. 1* (Oxford: Berg, 2004).

12 Kittler 10.

in Spiritualism, "the voice became the most important form of embodiment and manifestation for non-embodied entities: it was at once the most powerful and the most versatile form of witness to the unseen."[13] Media archaeologist Jussi Parikka addresses the psychosis that accompanied similar new media and "weird objects" of modernity, as having their own material logic. He channels Kittler, claiming that "the bodies of psychotics, or ghosts, is actually about technical media showing how they work as inscription-machines with the human body being merely a relay point for a wider discourse network."[14]

In recent years, as part of an attempt to better understand the nature of the poltergeist, psychical researchers have studied the molecular nature, sound signatures, and geophysical qualities of objects and environments affected by these apparent spirits. Several decades' worth of articles intent on providing proof-driven scientific data of the poltergeist have appeared in parapsychology journals, complete with images of sound waveforms gathered during research, presented as evidence. These cases are doubly rife for hijacking by object and sound theorists looking to apply methodologies in new ways. In an especially fascinating example from 2010, Barrie Colvin transformed 10 audio samples of the sound of purported poltergeist knockings into visual waveforms by using spectrographic computer software.[15] He found that the data patterns of a human-produced knock display a peaked threshold, whereas the apparent poltergeist knock showed a slow rise in volume, as if the knock had been produced from *inside* the object being manipulated. In effect, Colvin's research

---

13 Connor, Steven. *Dumbstruck: A Cultural History of Ventriloquism* (Oxford: Oxford University Press, 2000), 390.

14 Parikka, Jussi. *What Is Media Archaeology?* (Cambridge, UK: Polity Press, 2012), 57.

15 Colvin, Barrie. "The Acoustic Properties of Unexplained Rapping Sounds." *Journal of the Society for Psychical Research* (73.2 no. 899, 2010), 65–93.

attempts to demonstrate the absorption of the poltergeist into architectural space as a playback device. This circles back, once more, to the concept of the "stone tape." The poltergeist, as a "noisy spirit" is the most cut-and-dry example of the sonic spectre, where the concept of *the ghost* is absorbed and flattened into the tiny ontological space of *the material*, eager to manifest itself in order to be heard again.

Researchers based in the humanities who study the culture of the paranormal, as well as those who gather data as parapsychologists, have been using forensic methods to give objects and places voices long before contemporary fields of object theory were defined. Despite chuckles of ridicule from bona fide scientific communities, today's surviving psychical research laboratories have aimed to use highly controlled technical processes to dissect the root materialism and causal factors of physical manifestations and objects "touched" by the supernatural.

I don't purport to be an unwavering believer in any one subfield of object philosophy, or to steadfastly declare myself as part of a singular ontological movement in material or sound theory. Or even, perhaps to the frustration of some readers, to concretely position myself as either a believer or non-believer in the supernatural. Rather, I'm most interested in exploring the boundaries between the trifecta of material culture, sound, and belief. I value the voice of the object, the *thing*, above all else, and prefer to give space for its natural narratives to escape. In choosing to listen, as a subject, one almost becomes as ghostly *as the thing itself*. Such attitudes of study where one is forced to consider "non-normative" reality as having equal weight as concrete fact can generate discomfort and alienation within certain intellectual circles. To study things within the context of, say, speculative realism is much more acceptable than making a foray into the parapsychologist's laboratory. But the bridge between the two is certainly there, and it's one that we are about to cross. One is not supposed to wholly absorb oneself in or become "contaminated" by his or her research. But I am not the first, nor will I be the last.

I. DEAD LINES

ELECTRIC LINE EVENTS

The smell of rose perfume floating through the sulfurous tendrils of extinguished candles.[1] Cherry tobacco smoked in a pipe.[2] The sound of thick pencils. A noise like "sssssssssshhck." The belly of a snake sliding across rough paper.[3] She could hear that the line was continuous. The pencil never lifted from the page, and so there was no tapping of lead on paper on top of wood. They weren't making grocery lists down there. Sometimes, if things weren't flowing smoothly, they used a planchette to aid in their receptions. The wood heart carried the pencil in a hole, guided

---

1 D479.7. *Evil smells transformed into sweet fragrances, and vice versa.* ★Loomis White Magic 81; The Stith Thompson *Motif-Index of Folk-Literature* allows for the classification of folktale motifs across cultures. Rooted in the structural (some would say reductionist) approach to the study of folklore, this system implies that, while the content of stories may shift over time and geographic location, their function remains somewhat static. I have applied this framework onto the semi-autobiographical vignettes that appear throughout this book as a way to neutralize my own narrative against more fantastical stories, to heighten meaning, and to make jokes; see Thompson, Stith. *The Motif-Index of Folk-Literature* (Bloomington: Indiana University Press, 1955–58).

2 E555. *Dead man smokes pipe.* U.S.: Baughman; N. A. Indian (Seneca): Curtin-Hewitt RBAE XXXII 219 No. 41, (Teton): Dorsey AA o.s. II (1889) 150.

3 X251.1. *Why no weavers in hell. Devils annoyed at their noisy trade.* Flemish: DeMeyer FFC XXXVII 84 No. 27a.

by wheels, one of which was slightly off-track. Barely audible squeaks, in need of oil. Soft mumbling. Questions.[4] Have you ever heard a ghost walk?

The shapes were coming down the stairs, behind her, as they always did, every second Saturday night of the month.[5] She had learned, over time, to not be afraid of them. They were people once. "It's not the dead that will hurt you," her grandmother would tell her. "It's the living." Since she had been old enough to escape her crib, K. had snuck out of her upstairs bedroom, tiptoeing to the edge of the landing, following the worn carpet runner so as not to make the hardwood floor creak. She wanted to listen to what was happening downstairs.[6] Her clandestine movements were an unnecessary ritual, but she followed them anyway. She wasn't invited, yet *she* knew that *they* knew she was up there, listening. There were strangers in the house. Some of the voices were familiar. They spoke like they were from town—farmers with clipped Ontario accents. Other voices came and went, some with more exotic accents. Others were strained, ethereal.[7] She was sure they spoke in other languages. They bled out from the shadows. Straining her ears to listen through the murky dark was tiring, and so sometimes she would fall asleep, fetal form at the top of the stairs. She would wake up in her own bed in the morning, hazy-headed and disoriented. Nothing was ever said of her nocturnal wandering, but she would often pretend to have been sleepwalking, rather than risk being labelled as an eavesdropper.[8]

---

4 E557.1. *Ghost writes on wall the answers to problems of person in trouble.* U.S.: Baughman.

5 E421.4. *Ghosts as shadow.* U.S.: *Baughman.

6 D1412.3. *Flames draw person into them.* (Cf. D1271.) Jewish: Neuman.

7 D457.14.2. *Transformation: tongue to flame.* Jewish: Neuman.

8 E175. *Death thought sleep. Resuscitated person thinks he has been sleeping. He exclaims, "How long I have been asleep!"* Bolte 555; Wesselski Märchen 192.—India: *Thompson-Balys; Philippine: Dixon 235; N. A. Indian: *Thompson Tales 319 n. 154, (Calif).

In the morning, if she was awake before anyone else, she would search for clues. A slim envelope of money that had not been there the day before, hidden in the back of the bread bin—payment from the strangers and farmers who would sit at the familial ghost table.[9] The dining room table was the biggest and oldest piece of furniture in their house, and had "come over from the Old Country," according to her mother. They never ate at that table, and she wasn't allowed to work at it. Their family of five crowded around the Formica and chrome table in the kitchen instead. The dining room and its table were reserved for second Saturdays. Sometimes the adults didn't clean up from the night before, and she would find the tablecloth hand-tatted by her great-grandmother, with its hilly landscape of white wax candle drippings in the center, thrown over that special table.[10] Nothing else remained.

In family photo albums, she once found an image of her grandmother, back when she lived in the farmhouse, too. Scalloped white edges with a date: 1959. Her grandmother was sitting in an oversized armchair. Outside there were wisps of smoke from the burn-pile of leaves, which reflected through the warbled glass windows and onto the peeling botanical wallpaper.[11] The staircase was in the background, fading into darkness at the top. Her mother always said she "got a chill" going up those stairs. One night before bed, she told K. the previous owner had hanged himself there, "over a woman," on the fifth step.[12] Knowing this, she avoided the step altogether for the next few months, double-stepping up or down.

---

9 K1064. *Man dupes animals into turning their tongues upside down.* India: Thompson-Balys.

10 D1153.1. *Magic tablecloth.* *Type 569, 851, 853; BP I 464ff.; *Aarne MSFO XXV 118; Penzer I 25f.—Breton: Sebillot Incidents.

11 D1572. *Magic smoke carries power of spirits.* (Cf. D1271, V220.) Irish: Plummer clxvi, Cross.

12 E538.2. *Ghostly rope of suicide appears.* U.S.: Baughman.

As she grew older and the dining room continued to host second-Saturday nights, K. grew bold, edging her way down the stairs, one step at a time. The thick wood slats of the banister provided ample reconnaissance cover. She discovered that if she leaned forward while sitting on the fourth step down from the top of the stairs, daring to put her feet on the fifth step, she could see the back of her mother's head, sitting at the dining room table.[13] The table was round, but it was clear that her mother was the head of the table, ruling over the events. The patterns of this ritual were predictable, but the events were not. Candlelit faces edged out from her mother on either side, arms raised, elbows bent and hands clasped, enclosing her mother in the center of the circle. One matriarchal hand grasped a thick carpenter's pencil over crisp new paper. The other rested on top of a fat sheaf of bond at the left, ready to be snapped under her writing hand as the pages were filled.[14] Their eyes narrowed in the dark, just barely swaying.[15]

They did this while they waited for the pencil to begin moving, a half-dozen or so believers, waiting to ask their questions. Her aunt L. and uncle R. were always there, as well as her grandmother. The other faces she recognized from town, but there were strangers there too, people "from away."[16] The seats around the table were always full.[17]

---

13 E386.3. *Ghosts summoned by calling them by name.* Icelandic: *Boberg; England: Baughman.

14 F1034. *Person concealed in another's body.* *Penzer VII 114ff.

15 D457.11. *Transformation: eye to another object.*

16 C745. *Tabu: entertaining strangers.* Jewish: Neuman.

17 D1153. *Magic table.* *Types 563, 564; BP I 346ff.; *Aarne JSFO XXVII 1—96 passim, MSFO XXV 118; *Köhler-Bolte I 109; *Chauvin V 272 No. 154.—Irish myth: *Cross; Icelandic: Boberg; English: Wells 32; Swiss: Jegerlehner Oberwallis 297 No. 28.

WE ARE NOT YET EXPERTS
The Mimetic Dead, Paranormal Acoustics, and the Medium Jesse Shepard

On 9 September 1920, the former 18th president of the United States, Ulysses S. Grant, spoke the following words:

> You are now at the cross roads. Take the wrong turning and you will come to the skull and cross bones. I could say much more but we are not yet experts in this new mode of inter-communication and must be brief.[1]

The "we" that Grant refers to is comprised of an indeterminate number of spirits in the afterlife, spirits who have recently discovered a method through which to communicate with the living. Grant counts himself among this community of spectres because he belongs there. In 1920, Grant was dead, and had been so since 1885. This message was "recorded," as the preface of the book *Psycho Phone Messages* suggests, by a Los Angeles Spiritualist named Jesse Shepard, who wrote under the pseudonym Francis Grierson.[2] These spirits declared themselves to be unskilled in the "new mode" of communication. This was a highly dubious claim, considering that messages between the living and dead had been in regular exchange via the formal structures of Spiritualist mediums since the mid-19th century.

The "mode of inter-communication" used by the spirits was the Psycho Phone. Its operator was Jesse Shepard.

There are two forms of spiritual communication that will be referenced with regularity here: the resonating chamber of the séance room, and spontaneous sounds that are the outcome of purported hauntings. In the former, a recurring theme is that spirits and Spiritualist mediums alike, in the course of imitating speech or sound acts, or by acting as living soundboards for disembodied voices, are engaged in acts of mimesis. The living serve as copycat sounding boards for the dead. In the latter example, hauntings act as repetitious residues of the past, accompanied by the embodiment and reproduction of sensorial effects, projected by the *haunter* onto the *haunted*. Acoustic memories of the once-living disrupt the world of the everyday. Throughout the cultural history of hauntings, we often encounter narratives of apparent spirits reproducing sounds related to actions they made while alive: footsteps, whistles, laughter, snippets of conversation, and occasional sounds of distress. Technology has also impacted and complicated narratives of supernatural experience. Upgrades to technology are simultaneously absorbed into the narrative "flow" of the mimetic dead.

I. *Communication & Mortality*

Jesse Shepard's *Psycho Phone Messages* immediately calls to mind the trope of Thomas Edison's lost plans for a contraption that could contact the dead. Edison's invention of the phonograph inspired *Scientific American* to proclaim, "Speech has become, as it were, immortal."[3] But others have made ample mention (and misinterpretation) of Edison's intentions for such a device. In 1920, the Wizard of Menlo Park himself remarked:

I don't claim that our personalities pass on to another existence or sphere ... But I do claim that it is possible to construct an apparatus which will be so delicate that if there are personalities in another existence or sphere who wish to get in touch with us in this existence, this apparatus will at least give them a better opportunity to express themselves ... I merely state that I am giving the psychic investigators an apparatus which may help them in their work, just as optical experts have given the microscope to the medical world.[4]

As for Psycho Phone messages, Shepard's description is somewhat vague: "The psycho-phonic waves, by which the messages are imparted, are as definite as those received by wireless methods."[5] Shepard's "waves" are not processed through aluminum horns, through balloons made fat by the breath of mediums, or derived from chemicals that cause antennas to vibrate and telegraphic arms to tap out messages from the dead. Although these machines exist elsewhere,[6] invented by others, Shepard's case is simpler than it appears to be: The Psycho Phone is a metaphor. "Discoveries," Kittler notes, "frequently start with metaphors."[7] The Psycho Phone is Jesse Shepard himself, acting as a "direct voice" communicator, listening and relaying the urgent messages of past politicians, abolitionists, suffragists, other such socially engaged personas via psychographic or channeled writing from beyond the grave.

Shepard's concept of the Psycho Phone is further muddled by the existence of a motivational device with an identical name, invented by A.B. Saliger. The device consisted of a turntable which would be triggered by an alarm clock, and pre-recorded messages would penetrate (or annoyingly rouse) the sleeper's mind. According to an advertisement placed in *The New Yorker*, "this automatic suggestion machine enables you to direct the vast powers of your unconscious mind during sleep."[8] Early patents were issued in 1927 and 1928, with US Patent No. 1,886,358 granted on November 1, 1932 for an "Automatic Time-Controlled Suggestion Machine." A popular recording titled *Mating* projected Saliger's own pacifying voice over the horn of the player: "I desire an ideal mate. I radiate love. I have a fascinating and attractive personality. My conversation is interesting. My company is delightful..."[9] However, the absorption of the sounds of success while sleeping was only accessible to the wealthy. The Psycho Phone's price tag reached $235 in the lean years of the Depression.[10] Occasional discoveries of Saliger's machine in antique shops and listed on eBay auctions have produced misleading information among believers, leaving them certain they've exposed a physical example of Edison's lost etheric communicator.

Why did the supposed spirits channeled by Shepard feel the need to vocally indicate their trepidation with this not-so "new" technology? The first message Shepard claims to have received was from Speaker of the House and Republican leader Thomas B. Reed (1839–1902). On 7 September 1920, Reed-via-Shepard stated: "I am placing you in communication with some of the most far-reaching minds of the past hundred and fifty years. The psycho-phone is new and we are using it for the first time."[11] The giddiness over new technology in the spiritual world conjures images of a clandestine gathering of ghouls eavesdropping on an old-fashioned rural telephone party line, and Reed being bold enough to be the first to speak up. It's odd that the "spirits" are so flummoxed with the Psycho Phone, considering that the Spiritualist belief system was so scientifically saturated. The term "spiritual telegraph" was commonly used by 1921, including a mention in the title of *The Yorkshire Spiritual Telegraph*, the Spiritualist newspaper established in 1855.[12] Perhaps it was Shepard's own discomfort with the permeation of new technologies into everyday life that was to blame for the confusion, forcing him to adopt unknown terminologies.

Parallels have been drawn[13] between the fascination with the newly introduced telegraph and the operating modes of the "spiritual telegraph" of the séance chamber. As Connor notes:

> The séance in fact doubles, or ghosts, the structure of hapto-sonorous hallucination which made the technologies of the telephone, the gramophone, the radio, and the tape recorder, along with all their contemporary refinements, so comfortable and familiar from the outset.[14]

In her 1869 account of the history of Spiritualism, the medium Emma Hardinge Britten claimed that the version of telegraphy used in séances originated from instructions given by spirits.[15] Another strange device, simply labeled "the signaling instrument," was discovered in a file of unattributed documents at the archives of the American Society for Psychical Research in New York City.[16] The "instrument" was used to defraud séance attendees, with a medium placing a small telegraph in her bouffant coif, which was surreptitiously connected to a small wire that ran up to the ceiling and over a chandelier. The wire ended at the hand of a conspiring assistant, who would tap out Morse code on the head of the seated medium.

These examples form part of a "shadow world" of communications devices based on legitimate telephony technologies. Here, these devices have been reimagined for darker chambers and embedded with spiritual resonance. The shifting nature of mediumship could be viewed as a lineage of prosthetics evolving from analog into digital realms. The rise of "direct voice" communication (courtesy of the medium's hijacked vocal cords), to the sweep of a hand holding chalk to a slate. The language of telegraphy transformed into binary codes knocked out by spirits, to the use of electromagnetic field detectors by modern ghost hunters.

I. Dead Lines

The late Victorian paranoia of premature burial prompted Walther Rathenau to speculate, in his fictional short story "Resurrection Co.," that a telephone network could be connected to every coffin in a graveyard.[17] Unfortunately, rapid technological evolution, in combination with small production numbers of spiritual design solutions, resulted in the loss of surviving examples, in most cases. There is no distinct archive or museum to preserve the machinery of belief.

II. *Paranormal Music*

Direct interrogation of the dead is a noisy process. Séance chambers play host to rappings, resounding voices, strumming airborne guitars. Temporary possessions harness the vocal chords of the medium. Beside the acoustic manifestations of the séance room, there also exists a history of "paranormal music," a term paramusicologist Melvyn Willin defines as "examples of music where no physical sound source is apparent."[18] It's important to note, in this case, that the aural experience is in contrast to the *singular* sounds that accompany the typical haunting (footsteps, disembodied voices, knocking), but rather take the form of multi-instrumental or multi-vocal songs and orchestral arrangements that are performed with implied *purpose*. Marcello Bacci and Luciano Capitani claim to have captured heavenly choirs using a radio receiver in the 1920s, recordings of which are stored in the Institut für Grenzgebiete der Psychologie[19] The parapsychologist Scott Rogo received a letter from a woman named Elsie Haines detailing how, one night in 1921, she woke to the sounds of beautiful orchestral music, and became "enveloped" by it. She noted: "I realized it was coming from Heaven... We did not have a radio or even a Gramophone.".[20] Such reports of invisible choirs, or of the inner music experienced by mystics and saints, could be considered examples of an odd extension of Pierre Schaeffer's notion of sound without an originating source, known as acousmatic sound:

> Behind the acousmatic curtain, behind the presence of audio, one finds the technological apparatus in retrograde—disappearing into the void that sound represents, shedding some of its history, its presence, and traveling under the radar of media critique to be remediated...[21]

In this sense, sound issued from the most basic technology—the human vocal tract—is severed from its source. Abstracted and ungrounded, such noise can roam free as a potential catalyst for supernatural experience and, by extension, the conversion of potential believers.

There is also considerable documentation of music being detected at the deathbed. The psychical researcher Ernest Bozzano published a collection of these narratives, including an account of the death of Wolfgang Goethe. Members of the household accompanying his transition were mystified by the presence of music in the house:

> It's inexplicable! Since dawn yesterday a mysterious music has resounded from time to time, getting into our ears, our hearts, our bones. *At this very instant there resounded from above, as if they came from a higher world, sweet and prolonged chords of music which weakened little by little until they faded away* ... It was thus that the mysterious music went on making itself heard up until the moment when Goethe breathed out his last sigh...[22]

Sources of paranormal music can be attributed to devious causes as well. The 1668 case of the Tedworth Drummer, recounted in Joseph Glanvill's *Saducismus Triumphatus*[23] (1681), is an early publication that would, in effect, set the standards for future reportage on poltergeist events.[24] The problems began when Tedworth resident John Mompesson waged a lawsuit against an itinerant drummer, confiscating his most prized material possession—his drum. Soon after, Mompesson regretted his decision, as his house was now filled with vengeful nocturnal racket. Of course, even at this early point in the case, suspicions of fraudulent behavior were suggested. When Sir Christopher Wren visited Tedworth, he made the correlation between the ghostly drum beats exuding from the walls and the presence of a certain female servant.

The British Isles have long been the site of portentous narratives of ghostly pipers and drums in the sky. An apparent haunting of Cortachy Castle in Scotland was attributed to "a fine young drummer boy" who was caught in an adulterous act with the lady of the castle. "He was seized, boxed up in his own drum and flung off the highest tower to his death," according to John Brooks.[25] From that point on, he would beat his drum whenever there was a death in the family.

Other cases seem rooted to a "time slip" effect, also known as "residual haunting," in which tendrils of music from the past continue to bleed over into the present. These cases often involve spectral pianos and organs, as in the case of the Moberly-Jourdain incident. While visiting Petit Trianon at Versailles in 1910, two women believed they had fallen through a "time slip." In 1919, they published their experiences, under pseudonyms, as *An Adventure*,[26] claiming to have witnessed Marie Antoinette and heard paranormal music.

These cases all evolved from private events into publicized experience. The music was impermanent and went unrecorded by physical media. Therefore, the descriptions of the sonic qualities of each case developed after the fact, based on textual and oral documentation. Owing to the nature of this kind of delayed documentation, accounts of paranormal music fall as fast victims to the folkloric process known as "continuity and change." In this process, the transmission of memory adapts and shifts as new locations, interpretations, emotional responses, and mistakes become enfolded into the original story. Memory and the external recounting of an experience are no better than the telephone game played in grade school. As temporal distance lengthens, memorates enter the realm of pure legend.

Paranormal music is always described in relation to its spatial characteristics. Sounds are reported as faint, otherworldly, or omnidirectional. They are often from above, but may come from below, or be rooted to specific rooms of a dwelling. With the exception of music produced on instruments, the nature of sound in the séance room is consistently disembodied, without visible production. The transfiguration of sound particles into numinous experience has the potential to become an act of embodiment. The phenomena are alternately "taken in" to the bodies of believers—or deflected by skeptics. Believers interpret the sounds mimetically as spectral hands mimicking live musicians, producing proof (and comfort) in the potential for postmortem survival. Skeptics associate the sounds as being human hands mimicking the ghost, producing acts of fraud and allowing security in materialist values.

III. *Mechanical Mediumship*

Before physical mediumship fell out of vogue, the instruments of choice for mid-19th century ghosts seemed to be guitars, accordions, violins, and trumpets. The famous Scottish medium Daniel Dunglas Home was skilled at coercing purported supernatural forces to produce music in the 1870s and 1880s. At one séance, the investigator (and professed believer) Sir William Crookes presented Home with a new accordion, hoping on some level to debunk Home's tricks. The accordion began in a cage, then it was "taken out and put in the hand of the next sitter, still continuing to play; and finally, after being returned to the cage it was clearly seen by the company... moving about with no one touching it."[27] Jesse Shepard occasionally performed his musical séances with the lid of his piano closed. Emma Hardinge Britten, a medium who was present during a musical manifestation by D.D. Home, wrote extensively about music and Spiritualism. "At times the piano on which my choir rehearsed to my playing was lifted bodily up in the air, obliging me to request the good invisibles to let us proceed with our practice."[28]

With the discovery of the wholesale séance supplier catalog, *Gambols with Ghosts* (and its public shaming by stage magicians), musical séance manifestations needed to evolve in order to survive. In mental mediumship, subjectivity is also amplified, making exposures of fraud more difficult. And so this new kind of musical medium could act as a conduit for the spirits of dead composers, allowing the likes of Beethoven, Mozart, and Liszt to harness their hands to complete unfinished symphonies. Leading Spiritualist musical mediums typically claimed no professional training as a musician, or asserted that their channeled music was beyond their own amateur abilities. Jesse Shepard, who would abandon the Psycho Phone for musical channels, *did* receive training in his youth, in spite of the rumors of divine inspiration. Shepard did nothing to quell the legend that became attached to his inspired playing.[29]

Some musical mediums "receive" performances that are "transmitted," meaning their hands are guided by the dead; others report visions of sheet music, and they transpose the notes to paper. Charles Tweedale claimed to have received the lost recipe for Stradivarius's violin varnish during a 1940s séance.[30] An article from a 1906 issue of *Annals of Psychic Science* explained the study of a M. George Aubert in detail. Aubert was taken to the Institut Général Psychologique in Paris, where he channeled a work by Mozart. This was no easy accomplishment, considering a phonograph was projecting a work by Verdi in his left ear and another machine played a Sellenick piece in the right ear. In another round of experiments, to prove the depth of his trance, a researcher "thrust a needle into the left hand without causing Aubert to wink or slacken by a comma the tempo of the symphony which he finished with a masterly chord."[31] And no discussion of musical mediumship would be complete without brief mention of the "unfinished symphonies" channeled by Rosemary Brown (1916–2001). Beginning with a "visitation" by Liszt when she was seven years old, Brown's mediumship lasted for a period of twenty years, reaching its climax in the 1970s.[32]

IV. *Bookends*

Jesse Shepard's timeline could be thought of as bookends that enclose a Spiritualist Renaissance. His birth year, 1848, was the same as the unofficial founding of Spiritualism in Hydesville, New York. His death in 1927 occurred amid a phase of psychical research that sought to combine science with higher consciousness, or to explore the boundaries of exceptional human experience and survival after death.

San Diegans lay claim to Jesse Shepard, due to the presence of his mansion, Villa Montezuma. In reality, he was a nomadic character with a complicated chronology. Shepard was born in England in September 1848 and moved with his immigrant parents to the American Midwest in March 1849. In 1867, he had a small amount of musical study. The following year, he performed at several recitals in major US cities. In 1871, Shepard traveled to Russia, where he discovered Eastern Mysticism and attended his first séance performances. He was intrigued by these events, and so the Czar of Russia's medium, General Jourafsky, agreed to school him in the art of raising spirits.[33] Shepard's manifestations of spiritual presence included singing demonstrations and piano concerts, wherein he allowed his body to be taken over by composers or exotic ancients. While travelling overseas, Shepard earned a respectable reputation as an inspired performer, which allowed him to insert himself among royalty, the extremely rich, and the creative illuminati of the time. His supporters paid him in compliments, as well as money and the material fineries that would eventually amass into a collection housed at Villa Montezuma.

By 1874, he had arrived back in the US. While in Vermont, he met Helena Blavatsky, the founder of Theosophy. He sang what he claimed were channeled songs for her, but Blavatsky soon discovered that Shepard deceived her, having learned the songs while travelling in Russia.[34] By 1880, he was divining the communications of Egyptian spirits in the borrowed séance parlor of Mrs. H.H. Crocker in Chicago.[35] Shepard arrived in San Diego in 1887, during the Great Boom years, by invitation of the High brothers, who built the luxurious Villa Montezuma for him. He soon began to lose interest in Spiritualism, however, and turned to writing. Not quite two years later, in 1889, he moved to France, where he would remain until 1913. In spite of Shepard's questionable relationship with Spiritualism, it's undeniable that he paid his dues as an author, with the publication of *Valley of Shadows* (1909) representing the climax of his literary career. Back in the United States, Shepard once again turned to Spiritualism and toured to demonstrate his "spontaneous inspirational compositions" as the "World Famous Mystic."[36] He published his last book, *Psycho Phone Messages*, in 1921, while living in Los Angeles. In 1927, he died at the age of 79, destitute, at his own benefit dinner, the events of which will be explored momentarily.

V. *Weird House of Ghosts, This*

On November 3, 1920, the abolitionist and clergyman Henry Ward Beecher (1813–1887) relayed the following psychophonic message to Shepard:

> What is causing so much crime? Not one, but many elements of decadence, all operating together, among which I can name rag, jazz, high balls, cabarets, free verse, neurotic art, sentimental optimism, cheap notions of progress, neutral sermons, automobilism, lack of child discipline, absence of fear among people under the age of forty—evils which you may apply to all English-speaking countries.[37]

It's interesting that Beecher/Shepard should be so conservative in spirit, especially given the medium's own experience of growing to maturity in a time of great technological and cultural change. Between 1886 and 1888, the town's population increased eightfold, from 5,000 to 40,000 residents.[38] The city was modernizing its infrastructure, and the largest contributing factor of growth was attributed to the recently linked Santa Fe rail line. With such a swell in the populace, the reciprocal perks of city life were to be expected; these years "were the most gaudy, wicked, and exciting in San Diego's history."[39] With a fledgling city in full swing, locals desired a more cosmopolitan art scene. As a result, Shepard was brought to the city in 1886. Here we find the origins of the Villa Montezuma.

Spiritualism was widespread across the country in the 1880s, and San Diego was not spared the zeal of alternative belief.

The brothers William and John High, who ran a successful produce business, were active believers in the community. Their profits, however, were soon being funneled—and, eventually, fully drained—into the maintenance of Jesse Shepard and his companion, Tonner. The High brothers had been swindled by Spiritualists in the past, but that did not prevent them from making their most grand donation to the cause, which took the form of Villa Montezuma, an opulent Queen Anne mansion located at 1925 K Street in the Sherman Heights neighborhood. Acting on "instructions" relayed from William High's deceased wife, the High brothers were soon paying the architectural design firm Comstock and Trotsche[40] to build the Villa as a form of memorial.[41] Of course, the spirit of High's wife was careful to relay specific details about the interior and furnishings to the exacting and exotic aesthetic preferences of Jesse Shepard, who would soon take up residence within it as San Diego's leading medium.

Historical records and local newspapers provide conflicting accounts as to whether Shepard's advantageous behavior toward the High brothers was born of malicious intent or if he intended to act in good faith. It's also questionable whether he was committed to providing the people of San Diego a palatial setting to experience supernatural marvels on a long-term basis, or whether he intended to flee, as he did, in the short two years' time it took to drain the brothers' finances. Sam High, nephew to the brothers of the same name, published a defamatory narration of the loss of his rightful family fortune in the *San Diego Union Tribune* in 1913 under the headline "Weird House of Ghosts This / Built by Spiritualist as Home of Spooks."[42] High states that "it was hypnotism and nothing else. That man had us so hypnotized that we would have done anything under heaven he told us to."[43]

Whether Shepard was a "genius," an advantageous charlatan, or a naive romantic is irrelevant here. The Villa stands as an embodied artifact within the landscape of San Diego. From conception to construction, from its architectural detailing to its interior design, the building is embedded with Shepard's own beliefs, as well as the beliefs of the Spiritualist community at large. The house is typical of the Queen Anne style: multiple stories with an asymmetrical floor plan, corner turrets, highly pitched roofs with multiple gables, a diverse array of window types, heavy ornamentation and harmonious paint colors, imbricated wall shingle patterns, Tudor-inspired half-timbering, allegorical Classical relief panels. The Villa also features a dramatic Arabesque dome on the south corner tower of the house, which is topped by a metal finial with a serpent wrapped around it. This motif was carried over to a grotesque water spout attached to the southeast corner of the roof. The room beneath the dome acted as Shepard's study and, by extension, served as the foundation site for his literary career. The interior of the house echoes the excess of exterior detail: darkly stained wood paneling and coffered ceilings, deeply embossed Lincrusta wallpaper, elaborate lighting fixtures, and columns at every turn.

The music room and conservatory dominate the majority of the ground floor. Hidden chambers and crawlspaces were discovered in the room, no doubt to aid in Shepard's séance effects. Windows at the ground level contain intricate stained glass with portraits of inspirational figures. Goethe,[44] Shakespeare, and Corneille represent literature. Raphael and Rubens personify fine art. (It is said that the beard of Rubens, in glass, continues to gray over time.) Beethoven and Mozart embody music. A massive piano with carved legs sits below a window depicting Sappho the poetess, flanked on either side by allegorical figures taken from John Milton. For Shepard, such sumptuous exoticism was not frivolous, but a necessity in the production of spectacle.

The multi-layered, hive-like arrangement of visual and tactile space acted as a containment zone on both acoustic and optical levels. As the visual field of Shepard and his guests was fractured, dominated, and deadened by the excessive filigree and acanthus leaves, the lush sounds of the channeled music were absorbed by Villa Montezuma's wool carpets, velvet upholstery, and honeycomb ceilings. The music room was a space where believers were primed for numinous experiences. Just as Queen Anne detailing denied the logic of Euclidian geometry elsewhere in the house, inhabitants of the music room became disoriented, and thus were mentally prepared for anomalous events.

Shepard enveloped himself in an interior as lush as any robber baron's home should be. But these design decisions also act as signifiers of belief. Whether Shepard's public persona as a voice box for the dead was fraudulent or genuine is inconsequential; Villa Montezuma was built as a manifestation site for the *potential* of mysterious creative forces. A serpentine finial at the roof was believed to channel occult energies into the house, and it's unquestionable that Shepard's inclusion of inspirational muses, which were rendered in glass and brightened by the persistent sunlight of San Diego, was a calculated decision.[45] In a symbiotic relationship, Shepard intended these architectural details built into the Villa to act as sites of power absorption by way of contact magic, and to amplify the creative process through proximity to such beacons. Over time, the story of Villa Montezuma became absorbed by the legend of Jesse Shepard, just as Shepard became absorbed by his own legend.

VI. *Supernatural Execution*

Shepard did not write directly of his mediumship techniques. In 1894, the Spiritualist journal *Light* published an account by Prince Adam Wisniewski:

> After having secured the most complete obscurity we placed ourselves in a circle around the medium, seated before the piano. Hardly were the first chords struck when we saw lights appearing at every corner of the room... The first piece played through Shepard was a fantasie of Thalberg's on the air from *Semiramide*. This is unpublished, as is all the music which is played by the spirits through Shepard. The second was a rhapsody for four hands, played by Liszt and Thalberg with astounding fire, a sonority truly grand, and a masterly interpretation. Notwithstanding this extra ordinarily complex technique, the harmony was admirable, and such as no one present had ever known paralleled, even by Liszt himself, whom I personally knew, and in whom passion and delicacy were united.[46]

Another account, equally poetic, described how Shepard:

> ... seemed to keep notes suspended in the air for minutes. Now and then he would make a shining vessel out of such a chord, and then he would begin to drip little drops of melody into it, until the Grail seemed to rise before your vision, luminous with blood-red rubies.[47]

While we are left with many florid accounts, Shepard's performances are essentially lost. He left no sheet music, and all attendees of his séances are now deceased. Early phonographs existed in 1886, but were not widespread, nor did Shepard seem to be one to embrace new technology. As a result, no recordings from the medium's era at Villa Montezuma exist. An equal, albeit strange, void exists in relation to his later Spiritualist "rebirth" period upon returning to the United States. Shepard's music exists only through memory and oral testimony transposed to print at the time, unfailingly from the perspective of believers in awe of his abilities. In 1875, the *Oakland Tribune* published a description of a Shepard musical séance at Tubbs Hotel:

> After playing a few selections on the piano with the lights all on, he had the piano removed and the gas turned off, leaving the room in total darkness. He sang "Annie Laurie," in imitation of Kate Hays; a Russian song and one in which he took bass and contralto parts together, accompanying himself on the piano. The performance concluded with a terrific storm of sounds on the piano, in which the barbaric music of the ancient Assyrians and Egyptians was imitated...[48]

It's likely that this "barbaric music" was Shepard's signature performance, *The Grand Egyptian March*. In this piece, known as one of his "impressionistic compositions," he would use only the piano and his voice to create realistic simulations of the sounds of battle. "[M]arching armies, trumpets, drums, tambourines, battle clashes and cannon booms" were all said to be mimicked in Shepard's séance room.[49] A common motif is the medium's apparent ability to manifest the sounds of multiple instruments, even if only a piano and one performer were present—or, in some cases, recreating the sounds of entire orchestras. In one 1914 session, Shepard "improvised on the sinking of the Titanic,"[50] creating sounds that would have resonated with an audience still reeling from the 1912 tragedy. His chords were also described, in what is now problematic language, as "haunting, monotonous, [and] primitive. It was as if the horns and drums of some African village had become civilized without losing their original weirdness—as if their uncouth noises had become miraculously transformed into genuine harmonies while still echoing the strife of primeval passions."[51]

The acoustic intensity of the séance room at Villa Montezuma was due in part to the production of notes that seemed to emanate from sources other than Shepard's piano. It seemed impossible that the medium could sing in two voices or play compositions constructed for four hands.[52] These examples are given in order to demonstrate that, alongside the most obvious relationship to mimesis, spatial characteristics are also a factor. As Connor notes, "All sound is an attempt to occupy space, to make oneself heard at the cost of others. Sound has power."[53] In spite of being bound to the keyboard, or to one spot as he sang an aria, Shepard created acoustic effects that in turn mimicked different spatializations. The dead imitated the living, simulating the sounds of their former lives—copycat productions of other times and other places. "[T]he music swelled and became strangely urgent—I felt there was an image that wanted to break through—a consciousness of some mighty presence—and all at once it was there: 'The Nile!'"[54] The charged sounds of battle issued from the séance room consequently mimicked another time period, and through the slippage of time, additional effects of space and distancing were added to the production.

Using an approximation of living speech, the dead mimic their former selves when channeled through a medium. Shepard claimed to act as an advocate for dead composers, surrendering his own agency in order to share lost, unpublished, and new compositions. As the standards of musical performance were embedded with acts of magic, the recital space became a mirror. The notes became an incantation and a ritualistic calling up—an invitation to reflect upon the dead, as well as a demonstrative site for the reinforcement of belief. The piano was raised from its hefty, earthbound presence to an intermediary tool used to unlock and tease forward evidence of the otherworldly. Shepard doubled as a spectre, both living and temporarily dead, resting within the liminal space of musical performance. His music was "weird and emotionally powerful." In whatever he played, he believed the impressions of "intrinsic spirituality could be imitated through music."[55]

VII. *"I Can't Believe We Live Here…"*

The time that passed between 1889, the year of Shepard's abandonment of San Diego for Paris, and 1971, the year the Villa was added to the National Register of Historic Places (Building #71000183), is littered with odd tales about the Villa. The house passed through several owners, and more stories of séances, buried treasure, and violent behavior became embedded in its local history. Certainly these events contribute to the mystique of the house, but none are quite as exceptional as Shepard's own narrative. In 1968, the house was purchased with the intention of transforming it into a house museum, which it became in 1972. It remained open until a sudden closure in 2006.[56]

The building now acts as a text, albeit in the form of a detective novel. In life, Shepard did not write directly of his séance activities at the Villa Montezuma, with the exception of one brief mention in a letter to General Grierson, his father, explaining that he did not make music "in the day time."[57] After his death, Shepard's life at the Villa has been maintained primarily through the intangible means of local legend and gossip—some accurate, some fictional. As a piece of architecture, Villa Montezuma is an oddity. Despite being maintained, it remains empty and unlived in, allowing it to slide into the trope of the "haunted house on the hill."

Rumors of Shepard's continued spectral presence in San Diego have continued well into the 2000s. The accounts are scattered across vernacular sources, including paranormal websites and message boards. While locals often report the appearance of "ghost orbs" in digital photographs, there are also sightings of a seemingly immortal Abyssinian cat dubbed "Psyche," as well as the shadow of a "hanged butler" in an upstairs window.[58] Shepard himself chooses to manifest in the manner that makes the most sense—acoustically. Reports of faint strains of organ music issuing from the Villa are common, even after musical instruments were cleared out during renovation. When the house operated as a museum, guests often claimed to experience private acoustic manifestations. "Deborah," a guest visiting the museum in 2003, stated the following:

> The youngest girl and I went to Jesse's bathroom to get some water and she got quiet and had this strange look on her face. After a moment she said she heard old organ music and a faint chorus above her. We told the Docent and he said there was an organ that wasn't working right upstairs above that area![59]

Just as Shepard once channeled the spectres of his own past, self-professed psychics living in San Diego claim to be in communication with Shepard. Local medium Bonnie Vent used channeling methods identical to the Psycho Phone (and identical to prior generations of spiritual communicators). Even from the grave, the dead-channel Shepard has used the living-channel Vent to champion his idea for the Psycho Phone, calling it out by name. Vent also claims that Shepard gave her the following message in 2002, a full 75 years after his death:

> I'm sure some people wonder if I am always at the Villa. The answer is no. I do enjoy coming to see the tours. ... I usually like to stand right behind the actor and pretend I am once again putting on my show. Sometimes they sense my presence and forget their lines.[60]

Ghost hunters claim to have captured electronic voice phenomena stating: *"I can't believe we live here."*[61]

VIII. *Shepard's Final Performance*

The January 18, 1914 cover of *The World*[62] shows a stylized drawing of a mature Shepard, piano keyboard underhand, eyes closed, in the throes of spiritual inspiration. In the undulating black cloud that Shepard and his piano emerge from, there are also the faces of séance attendees who watch the performance either in awe or scrutiny. Flowing outlines of spirits in transitional states also appear. The closest revenants appear as spectral white upon black, watching over the medium. They are seemingly aware that they are being called forth to demonstrate. As they retreat further away from the piano, they transition to reenact scenes of battle from their life, becoming less present and responsive to the environment they have been called into, as they dissolve into the ether (and into the white of the page). A byline above the image reads: "Psychic Pianist Paints Music-Picture of Your Moods."

The facts of Shepard's death are as theatrical as his mystical enactments. In 1927, at the age of 79, the medium's destitution was at a critical peak. Los Angeles supporters rallied around him, attending a benefit dinner in his honor at his home. In the course of the evening, he would demonstrate his "instantaneous compositions" for the guests in attendance, echoing the scene described above. After playing several improvisations:

> ...he turned to the instrument and announced that the next and last piece of the evening would be an Oriental improvisation, Egyptian in character. The piece was long, and when it seemed to be finished he sat perfectly still as if resting after the ordeal of this tremendous composition. He often did that, but it lasted too long and I went up to him – he was gone! His head was only slightly bent forward, as usual in playing, and his hands rested on the keys of the last chord he had touched.[63]

This was how, on May 29, 1927, Shepard's chronology came to an end.

If we are to approach the production of para-acoustic sound as an extension of flesh that has been deterritorialized from the hands from which it was produced, it is in the process of its free-floating "collisions, abrasions, impingements or minglings"[64] against objects, environments, human ears, and minds where the narratives of this book are created. The musical medium is an agent of liminality, as he plants one foot on the shores of life and the other on the banks of death. The pedigree of musical mediums—Jesse Shepard among them—created, through spiritualistic mediation, a complicated sonorous void within which an audience could temporarily (and safely) suspend themselves and float among the sounds of another time and place.

Musical mediums working in the vein of Jesse Shepard created highly controlled environments where sound was seemingly brought forth on command, for a sustained period of time. Similarly, the rappings of the poltergeist, while initially chaotic, could be trained to rap in response to human interrogation, much as two foundational figureheads of Spiritualism, Kate and Margaret Fox, did. In these cases, séance attendees and mischievous girls were mentally prepared to either suspend belief or brace themselves for extraordinary experiences, or to act as the medium and interrogate the spiritual subject. These examples are in contrast to the effects of the feedback loops of everyday sound issued from a residual haunting, as well as deathbed music, portentous pipers, and invisible choirs, whose impromptu recitals are unresponsive to human persuasion to perform on command.

Whether prompted or not, sounds of purportedly paranormal origin become confused with the boundaries of ourselves. But this is perhaps more so the case with noises of mundane origin. The sounds with which this book is concerned, expected or not, absorbed or deflected, live within us as parasites, embodying narratives to be shared as hopeful—or willingly blocked from memory as traumatic. The anomalous sounds of séance rooms or acoustic hauntings are embedded with excess narrative, waiting to be received by "imaginative ears" that seek to tap into sounds that belong to an embodied network of belief that could never be confined to notes and staff on paper.

Notes

1. Grierson, Francis. *Psycho-phone Messages* (Los Angeles, Calif: Austin Publishing company, 1921), 26.
2. For purposes of clarity, throughout this chapter I will use Jesse Shepard's birth name, as opposed to Francis Grierson—a pen name he began to use in 1896. Grierson was his mother's maiden name, and Shepard was christened as Benjamin Henry Jesse Francis Shepard.
3. Kittler, Friedrich A. *Gramophone, Film, Typewriter* (Stanford, Calif: Stanford University Press, 1999), 72.
4. Edison, Thomas A. and Dagobert D. Runes. *The Diary and Sundry Observations of Thomas Alva Edison* (New York: Philosophical Library, 1948), 205.
5. Grierson, 15.
6. See Braga, Newton C. *Electronic Projects from the Next Dimension: Paranormal Experiments for Hobbyists* (Boston: Newnes, 2001).
7. Kittler, 30.
8. "The Psycho Phone," *The New Yorker* (New York: New Yorker Magazine, inc., etc., 1933: Vol. 9), 13–14.
9. Ibid, 13–14.
10. Ibid, 14.
11. Grierson, 23.
12. Spence, Lewis, Leslie Shepard, and Nandor Fodor. *An Encyclopedia of Occultism: A Compendium of Information on the Occult Sciences, Occult Personalities, Psychic Science, Magic, Demonology, Spiritism, Mysticism and Metaphysics* (New York: University Books, 1960), 877.
13. See Sconce, Jeffrey, *Haunted Media: Electronic Presence from Telegraphy to Television*, Sterne, Jonathan, *The Audible Past: Cultural*

Origins of Sound Reproduction and Kittler, Friedrich, *Gramophone, Film, Typewriter.*
14. Connor, Steven. *Dumbstruck: A Cultural History of Ventriloquism* (Oxford: Oxford University Press, 2000), 393.
15. Davis, Erik. *Techgnosis: Myth, Magic + Mysticism in the Age of Information* (New York: Three River Press, 1998), 61.
16. In 2008, I conducted primary research at the American Society for Psychical Research. Much of this research was directed toward uncovering devices and tools for mediumship, used to both perpetuate and detect fraudulent acts in the séance room.
17. See Rathenau, Walther, *Impressionen* (Leipzig: S. Hirzel, 1902).
18. "Music and the Paranormal," Melvyn Willin. <www.hulford.co.uk/musicpa2.html> Accessed 30 January 2016.
19. Fischer, Andreas. *Okkulte Stimmen – Mediale Musik: Recordings of Unseen Intelligences 1905–2007.* (Berlin: Supposé, 2007), CD.
20. Rogo, D. Scott. *Nad: A Study of Some Unusual "other-World" Experiences* (New York: University Books, 1970), 43.
21. Dyson, Frances. *Sounding New Media: Immersion and Embodiment in the Arts and Culture* (Berkeley: University of California Press, 2009), 13.
22. Rogo 1970, 64–66.
23. Also known as: *A Blow at Modern Sadducism... To which is added, The Relation of the Fam'd Disturbance by the Drummer, in the House of Mr. John Mompesson.*
24. Finucane, Ronald C. *Appearances of the Dead: A Cultural History of Ghosts* (Buffalo, NY: Prometheus Books, 1984), 10.
25. Brooks, John A. *Britain's Haunted Heritage* (Norwich: Jarrold, 1990), 215–16.
26. See Moberly, C.A.E, Eleanor F. Jourdain, Edith Olivier, and J.W. Dunne. *An Adventure* (London: Faber and Faber, 1931).
27. Britten, Emma H. *Nineteenth Century Miracles; or Spirits and Their Work in Every Country of the Earth: A Complete Historical Compendium of the Great Movement Known As "Modern Spiritualism"* (Manchester: W. Britten, 1883), 147.
28. Britten, Emma H., and Margaret Wilkinson. *Autobiography of Emma Hardinge Britten* (Manchester: John Heywood, 1900), 53.
29. Simonson 1966, 25.
30. Willin, 56.
31. "Echoes & News: The Musical Medium, Aubert" *The Annals of Psychical Science* (London: Office of the Annals, 1905), 129–131.

32. Brown's compositions numbered over 600 by 1965. As the compositions were analyzed by professional musicians and historians, the obvious differences in quality between her recordings and sheet music and the work of the deceased composers she claimed to be channeling caused the examiners to denounce Brown's claims as dubious.
33. Crane, Clare. "Jesse Shepard and the Villa Montezuma." *The Journal of San Diego History* 16.3 (1970). <www.sandiegohistory.org/journal/70summer/shepard.htm> Accessed 30 January 2016.
34. Simonson, 33.
35. Simonson, 74.
36. Davis, Erik and Michael Rauner. *The Visionary State: A Journey Through California's Spiritual Landscape* (San Francisco: Chronicle Books, 2006), 35.
37. Grierson, 72–73.
38. Crane, Clare. "The Villa Montezuma as a Product of its Time." *The Journal of San Diego History* 33.2-3(1987). <www.sandiegohistory.org/journal/87spring/villa.htm> Accessed 30 January 2016.
39. Pourade, Richard F. *The Glory Years* (San Diego: Union-Tribune Pub. Co, 1964), 193.
40. Crane, Clare. "The Villa Montezuma as a Product of its Time." *The Journal of San Diego History* 33.2-3(1987). <www.sandiegohistory.org/journal/87spring/villa.htm> Accessed 30 January 2016.
41. Simonson, 37.
42. *The San Diego Union Tribune,* (San Diego, Calif: Union-Tribune Pub. Co., July 20, 1913), 1.
43. Ibid, 29.
44. Though no record of it has been found at the time of this writing, it would be interesting to know if Shepard was aware of the narratives concerning the paranormal music reported at the deathbed of Goethe.
45. Local legend maintains that the finial, which functions as a lightning rod, is strange, owing to the fact that thunderstorms in Southern California are rare. Many believe it to be a device used to channel occult energy into the house.
46. Cited in Willin, Melvyn *Music, Witchcraft and the Paranormal* (Ely: Melrose, 2005), 55.
47. Bjorkman, Edwin. "The Music of Francis Grierson," *Harper's Weekly*, LVIII (February 14, 1914), 15.
48. *San Diego Union*, October 2, 1887, 22.
49. *San Diego Union*, June 12, 1887, 7.

50. Bjorkman, 15.
51. Ibid, 15.
52. Willin, 55.
53. Connor, Steven. "Ears have walls: on hearing art" (2005). *Sound*, ed. Caleb Kelly (London: Whitechapel Gallery, 2011), 136.
54. Bjorkman, 15.
55. Ibid, 14.
56. "The Villa Montezuma: History of a House." *Friends of the Villa Montezuma*. n.d. <www.villamontezumamuseum.org/Villa_Site/About_us-Main_pg.html> Accessed 30 January 2016.
57. Simonson 1966, 37.
58. "Montezuma Mansion." *Weird California*. 2006. <www.weirdca.com/location.php?location=64> Accessed 30 January 2016.
59. "Organ Music." *San Diego Paranormal Research*. n.d. <www.sdparanormal.com/Villa_Montezuma.html> Accessed 30 January 2016.
60. Vent, Bonnie. "Jesse Shepard Channel: 12/07/2002." *San Diego Paranormal Research*. n.d. <www.sdparanormal.com/articles/article/274216/1670.htm> Accessed 30 January 2016.
61. Davis 2006, 35.
62. Image appears in Clare Crane. "Jesse Shepard and the Spark of Genius." *The Journal of San Diego History*. 33.2 (1987).<www.sandiegohistory.org/journal/87spring/shepard.htm> Accessed 30 January 2016.
63. Cited in Simonson 1966, 137.
64. Connor 2011, 135.

ELIZABETH STREET APPORTS

The first afternoon when they entered the Elizabeth Street house, it was K., her mother, and her grandmother. K's mother had been given the key by her boyfriend the night before he left to return to his teaching job in Northern Ontario. In K.'s memory, he never once came into that house while they lived there, and no one had been inside the house for at least five years.[1] This is important to remember. The house was unnaturally unruly. The smell that poured over them when they stood in the entry foyer was not as offensive as it could have been. The air was stale with subtle undertones of sickness. Past the foyer, straight ahead, they entered a living room. To the right was a bright red door. This was the first portal they were compelled to open, beckoned by its emergency hue.[2] Behind the door was a closet containing the angled backside of the staircase. It was quite cavernous, the size of a small bedroom. K.'s mother pulled a cotton string to trigger the hanging bulb overhead. This room, compared to others, smelled fresh. The smell of sweet dough and eggs and butter wafted toward the trio. There were two

---

1 E593.3. *If no lamp is lighted in a house for a period of fourteen days, ghosts take it for their dwelling.* India: Thompson-Balys.

2 E599.11. *Locked doors open at touch of ghosts.* India: Thompson-Balys.

card tables set up alongside one wall. Their tops were lined with baked goods.

Nothing was moldy. K.'s mother reached out without pause and ran her finger through the frosting on a white cake, stopping short at bringing it to her lips. Everything was beautifully arranged. Artificial flowers surrounded platters of cookies and tiered cakes on polished silver risers. The cake's icing was fresh, like it had been spackled earlier in the day. K's mother touched a pyramid of chocolate cookies and jumped back in shock, declaring they were warm to the touch. The prospect of eating all of these desserts excited 8-year-old K. But her mother backed out of the room with a look of confused horror. She forced K. and her grandmother to follow. K. found this upsetting. What were they going to do about all of those treats? "Throw them out," her mother said, definitely. She was convinced they were a trap, a test laid out by a vengeful spirit in waiting.[3] There could be no other explanation for their freshness. The contents on both tables—pounds and pounds of hand-baked goods—were some of the first things that went into black trash bags and out onto the curb. They lived in the house for six months, but it felt much longer. K. often wondered if the house would have acted more amicably if they had partaken in its offerings.[4]

---

3 F473.6.3. *Spirit leaves food in table or cupboard.* England, Wales: *Baughman.

4 Z13.4 (j) >> (Man chased by coffin, which follows him).

II. UP THERE

NOTHING IS STATIC IN THE TIDES
From the Sea To Mars With Charles Francis Jenkins

I. *Stand by for Weather Map*

Descending from a flight over Portland, Maine via Detroit, I see spotty islands and stacked heaps of stone blocks forming cliffs below. Boats slice through the butter sea, raising triangles of white foam in their wake. The plane takes a sharp bank north, angling toward a half-round castle structure, secluded on its own island—a prison, a sanctuary, a quarantine bunker—or perhaps just a garrison fort whose presence was rendered ominous by its utter inaccessibility. Fort Gorges, as I would later come to know it, was in fact part of a triangulated defense line active during the American Civil War, alongside Fort Preble and Fort Scammel—a sod-roofed sneak camouflaging itself as a hillside. Likewise, these two stone guardians of Casco Bay emerge into view as the commuter plane lurches through a patch of turbulence and aligns itself with the landing strip.

The remote nature of these obsolete military strongholds recalls a photograph I had recently found of the laboratory and home of Charles Francis Jenkins, best known as an underdog pioneer in the development of motion picture and television technology. An aerial view of Jenkins's property, five miles north of Washington, DC, shows a flat, sweeping field bisected by a dirt road. On one side of the lane is a domesticated gabled house; on the other, a simple building, houselike, but clear in its utilitarian intent. The building is locked into its own kind of triangulation, solitary and yet an essential part of a whole, anchored between two soaring communications towers. This structure is the Jenkins Radiomovie Broadcast Station (W3XK), the first station to be granted a commercial television license,[1] in 1928. It was a modest hub, but Jenkins had grand designs to broadcast outward and into the world. After an especially vicious Midwest winter, the onset of a rainy spring has had me thinking a lot about Jenkins's lesser-known work on weather map technology, as well as radio waves and the sea, and the poetic systems that lay between all of these things, in Jenkins's time and our own.

In 1890, Jenkins began to experiment with an apparatus that would become the Phantoscope motion picture projector. By 1895, Jenkins and his collaborator, Thomas Armat, began to publicly demonstrate an improved model.[2] Audiences were impressed, but bitter feuds broke out between the two inventors over ownership of the patent, ending in Armat burgling Jenkins's house to steal the prototype. When the dust settled, Jenkins was forced to walk away, and Armat sold his rights to Thomas Edison. The apparatus was rechristened as something living—the Vitascope—and Jenkins was forced to find his feet again.

In his autobiography, *The Boyhood of an Inventor*, Jenkins describes occasions when he "ran out of the house at the noise of awesome honking, to watch the wild geese flying over, darkening the sky with their numbers."[3] Skies heavy with geese migrating from one coast to another pushed a young and mobile Jenkins to explore the North and Southwest. With this walkabout complete in 1890, he settled into a job in Washington, DC, working as a secretary to Sumner Increase Kimball and his United States Life

Saving Service.⁴ As Jenkins transcribed reports on his typewriter at the USLSS, he was emotionally affected by their contents, which described the sea as conscious, and those who fought to calm its fury as heroic. Regarding the rescuers, he notes "lonely patrols in blizzard weather" and tales "of sailors lashed to the rigging and drenched with freezing spray, or snatched overboard by the angry sea; and which often later laid the dead body on the beach, gently and tenderly, as though in atonement for her ungovernable fury of the day before."⁵

Within a year, Jenkins left the USLSS in order to pursue his mechanical experiments with motion pictures as a full-time career. But the magical qualities of wet weather systems continued to haunt him, even in the desert. In a letter to a friend, Jenkins wrote of his adventures camping in the "sandy waste" of Diablo Canyon, near Santa Fe, New Mexico in 1899. He relayed stories of distress caused by broken compasses, of silence while sleeping in a nest of sand. His travels led him to the Hopi Snake Dance ceremony, which he filmed using one of his experimental cameras. "The whole is an elaborate prayer for rain, and rain it did that night," he wrote in his autobiography. "It poured. A dry sink near the middle trading post was now a lake around which I must detour."⁶

In September 1923, Jenkins found himself, by way of invitation, among a party of naval officers aboard the *USS St. Mihiel* to observe "the experiment of sinking a battleship by airplane bombs."⁷ Several missed targets later, the 10,000-pound TNT bomb finally struck the mark, anchored a safe distance away. He and the crowd watched as the ship was reduced to sinking cinders. Later, he wrote to his wife, "We steamed over her grave in a great oil slick which marked the spot. ... Those of us who watched her go down were silent and rather awed, I think. ... It didn't seem quite sportsmanlike. ... I had troubled dreams last night, dreams of again seeing the bombs strike."⁸

Jenkins found his second life calling in 1925 as the first American to demonstrate an invention that could transmit moving pictures by "radiovision."[9] This same year, he was awarded a patent for "Transmitting Pictures Over Wireless," which we now more simply call *television*. Demonstrations were reported in the *Washington Sunday Star* in June 1925: "[T]he image of a small cross revolving in a beam of light ... while not clear-cut, was easily distinguishable."[10] The cross was in fact an image of a windmill spinning—a continuous four-foot loop of film that mimicked the motion of its contents, around and around, through the projector, for 10 minutes. The wavering image was relayed from the Jenkins Laboratory to the Navy's Anacostia radio station on June 13, 1925. This was the foundational moment when objects at a distance were first seen in this way.

The sea continued to find Jenkins. In July 1926, representatives from the Navy and the Weather Bureau, as well as a group of ship captains, gathered in Jenkins's laboratory to discuss the development of an experimental weather map service that would harness the visual nature radio facsimile as its backbone. Jenkins was a fast worker, and by August, a prototype was ready to test. The transmitter consisted of a motor-rotated glass cylinder, with a printed map wrapped around its circumference. The idea was that, hour by hour, on board the ship, "after the usual code announcement, the radio man would hear: 'Stand by for a weather map' ... In a few minutes a weather map in red ink on the brown printed based map would be ready to hand to the captain of the ship."[11] The first map was telecast in this way on August 18, 1926, and Jenkins became the father of marine facsimile.

A network of locations was outfitted with weather mapping instruments. Jenkins's rural laboratory remained the central observation site, with three additional dispatching devices sent to operators at the Navy Building and the Weather Bureau in

Washington, and one to Chicago. They were placed aboard the Atlantic fleet flagship, the *USS Trenton*, as well as the *USS Kittery*.[12] The ships cruised around hazardous sites in the Caribbean Sea that were known to cause major issues with static in on-deck communications devices.

Jenkins converted wireless radio into lifelines—beacons to keep ships buoyant against the volatile dangers of storms and fog. The speed of warning signals sent by earlier iterations of the telegraph could be slow. And messages ran the risk of becoming unreadable if static garbled the transmissions received by a ship's communication systems. Jenkins's weather maps, however, could be read for their context, on the whole. Weather data was transmitted as a picture, printed line by line over a standard map base. So even if interference stalled the map for a few lines, it could still be coaxed into comprehension. This was especially useful in static-producing storms at sea. In September of that same year, the *USS Kittery* was able to use the weather map system to change its course and avoid the "maximum fury" of the Great Miami Hurricane.

From September 1926 until March 1927, the collaboration between Jenkins Laboratories, the Navy, and the US Weather Bureau continued. This type of small-scale experimentation served as a proving ground, but there was no competing with the improved systems being offered by major telecommunications powerhouses like RCA. Jenkins soon shifted his focus to the mechanical scanning television. But the basic technology of his marine weather facsimile system remains a vital tool for anticipating the threat of seaborne storms, even today.

It is no stretch to believe Jenkins would have encountered a slew of old sailor superstitions in his time working with the Navy and the USLSS—and that he could have recognized truth in some of them. As he worked to close the gap between the qualities of weather forecasting on land and at sea, by means of channeling radio waves, the weather and its mysteries continued to fascinate Jenkins. Perhaps even more fascinating is the fact that he consequently rendered centuries of weather lore obsolete, especially that old saying "What the sea wants, the sea will have."

―――――

Mirages projected up into the clouds above the ocean—fata morgana—are tricks of the eye, dangers designed to lure sailors at sea to their demise. Spired castles packed with treasure. Ghost ships that always remain out of reach, shifting form and skewing like a misaligned vertical hold on a television screen. The potential for infinite reflections and diffractions in the low-hanging space beneath the water is a form of phantasmatic communication—mischief divorced from intent, long-distance calls made by the Earth to the sky. Here, the tides never retreat.

―――――

II. *Turn Me On, Dead Man*

October 12, 1969. Michigan fall, a blustery Sunday afternoon. All-day rain and fog. Broadcasting from WKNR-FM in Dearborn, just 10 miles outside Detroit, DJ "Uncle" Russ Gibb received a call-in from an anonymous student. The caller, identified only as "Tom," urged Gibb to play the Beatles' "Revolution No. 9" backwards, to partake in the garbled chant, "Turn me on, dead man." The rumor of Paul's death wasn't new—it had been circulating around college campuses for a while, and the underground media stacked their own loose theories on top of vague suspicions. Wildfire rumors repeated the incantation "Paul Is Dead" (and had been so since 1966). Paul—or perhaps his lookalike, "Faul"—made a public rebuttal in *Life* magazine. Superfans logged hours hovering over their turntables, teasing out secret codes from films, album covers, and the backwards spin of Beatles records. Lennon in white led the funeral procession across Abbey Road. A rogue force was killing the rock star while he remained alive. Sound waves are subject to transformation, to carry messages between the living and the dead. For media theorist Friedrich Kittler, "death is primarily a radio topic."[13]

As the American frontier closed, there was a need to explore further. In 1893, when historian Frederick Jackson Turner presented his "frontier thesis" at the World's Columbian Exposition in Chicago, he declared that the "primal American experience was over."[14] The line between civilization and the frontier had blurred. America had swallowed itself. The wilderness had shifted from land to ocean and from ocean to land. And so, in a post-borderland society, where else was there to go? Alien forms were evidently absent in a closed frontier, and so we turned upward to the sky once more.

The French *voyante* Hélène Smith talked to Mars long before it was fashionable to do so. Building her reputation among Spiritualist circles, Smith attained a degree of fame in the late 19th century with her automatic writing transmissions. Her communications were received by her familiar spirit, Leopold, who had been reincarnated many times. In the most popular messages, Smith relayed stories she claimed to receive from Leopold's visits to Mars, which she would translate from an advanced constructed language of symbols (deemed authentic by Ferdinand de Saussure, no less) back into French. Théodore Flournoy, her psychologist and friend, published *From India to the Planet Mars*, a book of transmissions grouped into various romances, and the "Martian" and "Ultramartian" cycles. Smith's "receptions" follow a chain of removal from the source that mimics media, like a radio transmission with increasing static.

The anxiety of coping with the vast and unknowable nature of outer space was tempered by the creation of earthbound technological devices that could mediate fantastical thoughts, attempt contact, and measure things we couldn't see. Science and technology became part of the quintessential American experience, sufficiently alienating—making strange—what had only just begun to seem familiar. The retrograde cycles of settlement and the primitive began to feed back upon themselves at a much faster pace.

The drive to eradicate superstition in modernity has led to skepticism. We accept, en masse, that we have lost our ability to deeply *know things* about magic. But the power of the ritual can be found lingering just below the surface of the multifarious histories of media, interlocking strata of meanings imprinted upon a vast archive of physical media formats and devices. When we record sound, we store time, archiving our own impermanence. When we are haunted by voices that stand outside of the exactly *here and now*, it is because we are being *touched from a distance*. With the knowledge of sound wave properties becoming widespread, inventions were created to tap into them, see them, manipulate them, volley them for distances.

Consider a less common story of the spectral media variety, one that falls somewhere between media rumor hype and my own personal experience: the history of Charles Francis Jenkins and Dr. David Todd's experiments to "listen to Mars" in 1924. The markings of the rectangular slab of wood and metal I'm peering at through the black netting of an industrial warehouse shelf tell me that it was made by the National Electric Supply Co. in Washington, DC, and that it is a Model SE 950. Its record tells me it was made on March 26, 1918, and that it was a Navy-type radio receiver used by Jenkins in his laboratory. It was designed to solve a problem—to be rugged and reliable on the battlefields of WWI—but this SE950 clearly saw no action. And so, with no immediate legibility of use, it seems to have carved out a comfortable niche as an abstracted "black box" of the Latourian variety, surrounded by other objects I "inherited" as a recently appointed curator at the Henry Ford Museum.

Charles Francis Jenkins was born north of Dayton, Ohio in 1867 to a Quaker family. Until the Great Depression hit, Jenkins had found some success with his invention of the mechanical scanning television in 1923, and was granted the first experimental television station license in the United States, in 1928.[15] In spite of being considered the underdog inventor of early television, Jenkins never referred to his invention as anything but "radio vision." He sold $7.50 "radiovisor" conversion kits to amateur radio enthusiasts (a price point less than the cost of manufacturing), in hopes that the viewership for his regular telecasts would continue to grow and be entertained. These telecasts, which were essentially tableaux of moving silhouetted figures, were crude in comparison to later television technology.

In 1924, as Mars drew close to Earth, the excitement of the American public grew considerably. The oppositional distance of 35,000,000 miles was a long shot, but was still close enough to study the planet and test observations made many years prior by astronomer Percival Lowell.[16] Lowell published three books about Mars and became the leading voice advocating that an advanced Martian society was plausible. The first book, published in 1895, explored the notion of Mars as an intelligent planet in an unprecedented degree of depth—illustrations and all. According to Lowell, the images of lines he saw through the eyepiece of his telescope appeared to be an intentional infrastructure. He spent a significant portion of his career promoting the idea that the lines were evidence of Martian-made irrigation canals crisscrossing the surface of the planet. A web of canal networks like these, like "network[s] of wires, turns a landscape, with all its jagged accident and lumpy irregularity, into an idea."[17]

As Mars' arrival loomed—an idea coming closer—the most unconventional scientists tallied up schemes to determine "how much water a problematical Martian would have for daily drinking, bathing and shaving."[18] In the end, the canals proved to be an optical illusion produced by a glitch of the observing telescope. Lowell was not the perpetrator of a purposeful hoax, but a victim of persistent visual misinformation. The imminent arrival of Mars and the potential for interaction were serious business. The Chief of Naval Operations, Edward W. Eberle, sent the following telegram on August 22nd:

> 7021 ALNAVSTA EIGHT NAVY DESIRES COOPERATE ASTRONOMERS WHO BELIEVE POSSIBLE THAT MARS MAY ATTEMPT COMMUNICATION BY RADIO WAVES WITH THIS PLANET WHILE THEY ARE NEAR TOGETHER THIS END ALL SHORE RADIO STATIONS WILL ESPECIALLY NOTE AND REPORT ANY ELECTRICAL PHENOMENON UNUSUAL CHARACTER AND WILL COVER AS WIDE BAND FREQUENCIES AS POSSIBLE FROM 2400 AUGUST TWENTY FIRST TO 2400 AUGUST TWENTY FOURTH WITHOUT INTERFERRING [sic.] WITH TRAFFIC 1800[19]

Jenkins's "radio photo message continuous transmission machine"—or, more simply, the Jenkins Radio Camera—was to be put to task for the mission. The apparatus had the ability to picturize sound produced by radio phenomena, producing a long strip of photographic paper that could be developed to reveal the chatter of sound waves.[20] "As the film unwinds, an instrument passes over it from left to right fifty times every inch. An incoming sound causes a light to be flashed on the film, and this produces a black line."[21] Over the course of three days, the Army and Navy silenced all radio transmission for short periods in hopes that any errant spikes on Jenkins's machine would detect signals from Mars. Advertisements in radio magazines requested that enthusiasts cooperate with the radio silence, but many proceeded to build and tune their own specialized sets to monitor Mars, hoping to record their own reports of unknown origin.

Doubtful that his Radio Camera would be useful to the cause, Jenkins nonetheless agreed to work alongside Dr. David Todd, the former head of Astronomy at Amherst College, and William F. Friedman, the chief decoding expert from the Army's Signal Intelligence Service. In the event that communications were received, Friedman was brought on board to be a codebreaker to interpret what would surely be a complex language or mathematical enigma. Todd clearly believed in the gadget's capacity for success:

> The Jenkins machine is perhaps the hypothetical Martians' best chance of making themselves known to earth. If they have, as well they may, a machine that now is transmitting earthward their 'close-up' of faces, scenes, buildings, landscapes and what not, their sunlight values having been converted into electric values before projection earthward, all these would surely register on the weirdly unique little mechanism.[22]

In preparation for the experiment, the US Naval Observatory sent an antenna 3,000 meters above ground in a dirigible pointed toward Mars. If anything interrupted the silence, the signals would be broadcast back to the Jenkins Laboratory's receiver, which would in turn relay the triggering data to the Radio

Camera. Recording commenced on August 21st for five minutes, at the top of every hour. For three days, everyone simply listened and waited.

When the film was developed 36 hours later, the American public was surprised when the print contained "messages" of dots and dashes and "a crudely drawn face," images that repeated themselves like a heartbeat down the entire 30-foot length of film. Jenkins, who had counted on the film being blank, refused to admit that the signals were anything but a scientific curiosity divorced from any potential that the communications received were indeed from Mars. Instead, explanations for the signals ranged from the static of a trolley passing by on the hour to misaligned equipment to the natural symphonic radio waves produced by Jupiter. In the end, the static discharge of those passing trolleys muddled the results into a perfect canvas for the public to project their desires about the possibility of extraterrestrial graphical communication. Jenkins feared that his machine would be perpetrated as a hoax and grouped with all other manner of pseudo-scientific apparatuses. Moreover, he worried that his reputation would be tarnished, so he released the films with a caveat: "Quite likely the sounds recorded are the result of heterodyning, or interference of radio signals. The film shows a repetition, at intervals of about a half hour, of what appears to be a man's face. It's a freak which we can't explain."[23]

In her study of soundscapes of the early 20th century, Emily Thompson explores how the "rise of electroacoustic devices redescribed sounds as signals, which allowed for the measurement and standardization of soundscapes."[24] She goes on to talk of the sounds of one's time as "an auditory or aural landscape ... simultaneously a physical environment and a way of perceiving that environment; it is both a world and a culture constructed to make sense of that world."[25] This is an important link that applies to Jenkins and his Radio Camera. This apparatus was similarly capable of capturing the graphical qualities of signal frequencies, allowing people to experience sound *through vision*, rather than as a fleeting auditory event, in the moment.

Today, the graphical score to such events can be codified back into sound again by using freeware computer applications like Nicolas Fournel's *Audiopaint*. The field of spectrographic sound recovery is a useful method that has practical and artistic applications. Patrick Feaster defines "educed audio" as a process to "bring out, elicit, develop, from a condition of latent, rudimentary, or merely potential existence."[26] The "intelligence" of the static read-outs that were produced in the Mars experiments are perfectly recoverable, using a variety of processes related to "paleospectrophony."[27] If Jenkins's SE950 radio were to be powered on and attached to an amplifier, the "Martian static" could be re-territorialized back into space. The radio's operation wouldn't change at all from the last time it had been used in the 1930s, save for the fact that a large audience would be able to hear the audio, rather than just Jenkins and Dr. Todd, who were the only people present during its original capture.

The most active points of Jenkins's career occurred during a time when the application of technology was hopeful, before it became a measure of contemporary anxiety in the postwar and Cold War era. And while Jenkins operated in a *physical* sense from his laboratory just outside of Washington, DC, it was his ability to navigate and tame the mystifying *invisible* landscape of wireless signals, creating new mechanisms for their interpretation, that is most enchanting. Jenkins became a poetic conduit for weather systems, Martian communications, and the then-unknown electrical impulses that could be transformed into images and sound. He tamed the sublime space of electronic and, by extension, spatial frontiers. By providing translation devices, he created carriers for new techno-sensual experiences in domestic space (like "radio vision visors") and demonstrated how one could transform mediums by fine-tuning radio frequencies to send "pictures through the air."[28]

The blurry boundaries of the radio cause it to behave like an object that seems like it should "do more, and mean more" than it lets on.[29] It is a double-dealing thing, a pliable thing, something that can be used for conjuring, but ultimately it is a *non-compliant* thing. The Jenkins experimental radio was designed as a military-use receiver, though that became secondary to its eventual employment as a machine to capture extraterrestrial transmissions. Yet the connection Jenkins's radio was expected to make was dubious from the start; even if it were to have somehow succeeded in its ability to "listen to Mars," it would remain subject to doubt. Jenkins expected radio to act in a way that was against its nature—to reverse its ear and listen, to act as part of the Earth's nervous system, to become a detector of aberrations in the sky.

Hannah Arendt referred to the rise of space consciousness in the 1950s as "worldly alienation."[30] It isn't too far a leap to connect Jenkins's own nascent alienation (minus any sense of paranoia) with the rise of space consciousness in popular culture in the 1950s–60s, and to the establishment of the SETI Institute, the extraterrestrial listening organization, in the late 1990s. Ultimately, Jenkins's belief that it was possible to scrub the airwaves back to their uncluttered, tabula rasa beginnings was a folly. When Jenkins and Todd tapped into the possibility of extraterrestrial communication, they were divided between skepticism and hope, though they were united in their curiosity.

The device from the Henry Ford Museum's warehouse is in fact the receiving radio from the August 1924 experiment, otherwise known as the device that triggered the Photo Radio printer. The sound waves captured during the experiment are visual records of a piece of technology pushed beyond its limits. A collision of various phenomena, both manmade and natural, forced the radio to record and to talk back. But the squeaks, yowls, and squawks of early radio also had a visual counterpart. Static and noise can produce apophenia, the desire to see things where there is no thing—images buried in the sparks. Or the feeling of being haunted from the deepest point of a record groove. Nothing is *really* static.

Notes

1. National Capital Park and Planning Commission, *Maryland Historic Sites Survey (M:31-10)*, "Jenkins Broadcasting Station," Maryland Historical Trust, August 21, 1975.
2. Lubachez, Ben J. *The American Cinematographer*, "The Beginnings of the Cinema: Birth of the Final Form of the Motion Pictures—The Work of C. Francis Jenkins" (Hollywood, Calif: ASC Holding Corp, May 1, 1922), 19.
3. Jenkins, C.F. *The Boyhood of an Inventor* (Washington, DC: Printed by National Capital Press, Inc., 1931), 5.
4. Jenkins, 67; Kimball was the first to unify local seafaring rescue organizations, an effort that would transform into the US Coast Guard in 1915.
5. Jenkins, 68.
6. Jenkins, 96.
7. Ibid, 143.
8. Ibid, 143.
9. "First Motion Pictures Transmitted by Radio Are Shown in Capital," *Washington Post*, (Washington, DC: June 14, 1925), 1, 3–4.
10. "Radio Vision Shown First Time in History by Capital Inventor," *Sunday Star* (Washington, DC: June 14, 1925), 1.
11. Jenkins, 156.
12. Ibid, 158.
13. Kittler, Friedrich A. *Gramophone, Film, Typewriter* (Stanford, Calif: Stanford University Press, 1999), 3.
14. Edgerton, Gary R. *The Columbia History of American Television* (New York: Columbia University Press, 2007), 43.
15. Edgerton, 30.

16. Connor, Steven. *Paraphernalia: The Curious Lives of Magical Things* (London: Profile Books, 2011), 152.
17. Connor, 152.
18. "Mars' Latest Visit Leaves Riddle of Life Unsolved: Red Planet Probed by Watchers in Search for Clues That Life Exists on Its Surface," *Popular Mechanics Magazine* (New York: Hearst Corp., January 1927), 8–12.
19. Telegram from the Secretary of the Navy to All Naval Stations Regarding Mars, August 22, 1924, Record Group 181, Records of Naval Districts and Shore Establishments, 1784-2000, ARC Identifier 596070, National Archives and Records Administration.
20. "Weird 'Radio Signal' Film Deepens Mystery of Mars: Pictorially Recorded Messages Here Mere Tangles Mass of Dots and Dashes—Growing Wonderment May Bring Tenable Interpretation Theory," *Washington Post* (August 27, 1924), 3.
21. Ibid, 3.
22. Ibid, 3; Todd was an oddball character with his own story. His wife, Mabel Loomis Todd, had a longtime affair with William Austin Dickinson, brother to poet Emily. The Todd Crater on *Phobos*, a satellite of Mars, is named after him, as are *511 Davida* (the sixth-largest asteroid) and *510 Mabella* (a minor planet that circles the Sun). The astronomer apparently suffered from erratic behavior, leading to a gradual retirement from the Amherst faculty and a series of institutionalizations late in life.
23. Ibid, 4.
24. Helmreich, Stefan. "An Anthropologist Underwater: Immersive Soundscapes, Submarine Cyborgs, and Transductive Ethnography." *The Sound Studies Reader,* ed. Jonathan Sterne (New York: Routledge, 2012), 171.
25. Thompson, Emily A. *The Soundscape of Modernity: Architectural Acoustics and the Culture of Listening in America, 1900–1933* (Cambridge, Mass: MIT Press, 2002), 1.
26. Feaster, Patrick. *Pictures of Sound: One Thousand Years of Educed Audio: 980-1980* (Dust to Digital, 2012), 49.
27. Feaster, 29.
28. "Moving Pictures Sent by Radio," *Popular Mechanics* (Chicago, Ill: Popular Mechanics Co, September 1925), 408.
29. Connor, 3.
30. Crowley, David, and Jane Pavitt. *Cold War Modern: Design 1945–1970* (London: V & A, 2008), 165.

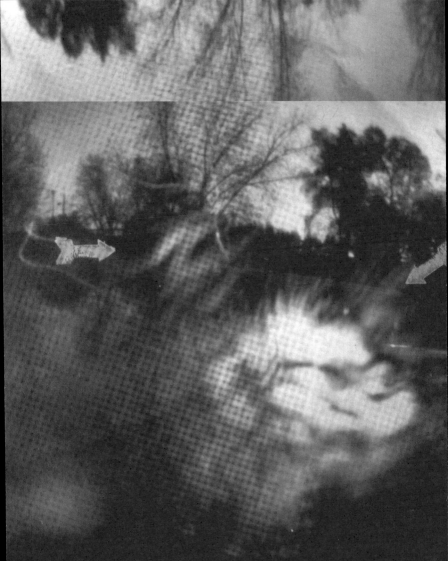

BOOK STREET SKIES

It was August, and K. was very young. Small enough to be sitting in a stroller on her grandparents' front porch.[1] No one in her family had air-conditioning. On the days when no amount of relief could be directed into the house by their rattling box fans, they sat on the porch with sweating mugs of pink lemonade.[2] Sometimes they waited for thunderstorms to break the humidity. The rumble of an oncoming storm signaled a weather event that could "get away from itself," as her grandfather would say. They stayed outdoors as the rain started to pound down.[3] K. in her fabric stroller sling, flanked by her grandparents and an uncle leaning against the porch railing, smoking, exhaling upward. K. remembers the sharp smell of dirt and sulfur and chlorophyll in the air as it mixed with tobacco smoke.

The Libbey Glass Company was a few blocks away, its massive cooling tower just visible from the porch. (At least the

---

1 C735.2.9. *Tabu: to rest sitting or lying until answer to certain question is learned.* Irish myth: Cross.

2 L144.2. *Farmer surpasses astronomer and doctor in predicting weather and choosing food.* Lithuanian: Balys Index No. 2448*; Russian: Andrejev No. 2132; Rumanian: Schullerus FFC LXXVIII No. 921 II*.

3 A287.0.1. *Rain-god and wind-god brought back in order to make liveable weather. Have been banished by sun-god.* India: Thompson-Balys.

tower *seemed* colossal in a town where nothing was over three stories tall.) Her grandfather had recently retired from Libbey, having worked for the company since childhood, when he needed a box to reach the assembly line. It was a Sunday night, and the factory whistles were silent. There was no industrial clamor and the lights were off on the exterior of the plant.[4]

Everything that happened next unfolded in *front* of the clouds. A ball of lightning appeared and hovered above the tower, as if a celestial light switch had been flipped.[5] It was very abruptly *there*, crackling. A sound that may have been real or misremembered backfill. The visual impression of a firefly the size of an automobile will always remain questionable in K.'s memory. The singular ball of light split into smaller segments and sped down into the ground, trailed by a rain of sparks.[6] Everyone on the porch was gaping, pointing. K., in her stroller, was pointing too. No one was talking, only silently gesturing. No one was talking later, either. The telephone wires in the four-block radius around the glass factory were fried to a crisp as the ball of light made its escape into the earth.[7]

---

4 C401.4. *Tabu: speaking while raising sunken church bell.* See all references to †V115.1.3.1.—England, U.S.: Baughman.

5 A1141.7.1. *Lightning as torches of invisible dancers.* Africa (Fang): Trilles 174.

6 F101.6.2. *Escape from lower world on horse of lightning.* India: Thompson-Balys.

7 A2218.8. *Eel burned by torch: hence red eyes.* Marquesas: Handy 80.

1890

Ethan Squier

George O. Squier

Mary A. Squier

Millie Slate

Belle Atwell

Minnie Harris

SQUIER'S MUSIC FOR THE TREES

Walter Van Dyke Bingham, a professor at Carnegie Institute of Technology and the author of *Studies in Melody*, published in 1901, was contracted by Thomas Edison in the 1920s to conduct studies on the effects of music on the moods of human beings. These "mood tests" were essentially about using sound for the purposes of stimulation—stimulation of desire and, once "background music" entered industrial settings, stimulation of efficiency. Bingham's forays into the applied science of sound were used to investigate the physical and psychological effects of several hundred selections of music played on an Edison phonograph—songs noted to have emotional impact upon the listener. Bingham's study about the benefits of listening to music within the domestic sphere, *Mood Music*, then became the basis of a marketing campaign for phonograph dealers to sell customers on the idea of holding social "mood changing parties" to modify the mental states of guests through the use of sound.

A 1921 advertisement for "The New Edison, The Phonograph with a Soul" from the James Hislop Co. asks "Will You Join Mr. Edison in an Experiment?" The text sets the scene for the reader: "Imagine you have just come home from shopping. You are tired and nervous. You step to the New Edison and put on an Edison Re-Creation. Gradually the music soothes you. You forget fatigue and your 'nerves' disappear."[1] This experiment in "applied science" about the idealistic effects of music was crowdsourced, inviting consumers to be of service to Edison directly: "[A]sk us for a supply of Charts and invite your friends for a Mood Change Chart party. They will find it more entertaining than the Ouija board. ... The New Edison has perfect realism."[2]

Edison and Bingham weren't alone in thinking about the power of sonic control to influence human moods and to manage the flow of tasks. Toward the end of World War I, George Owen Squier, the Chief Signal Officer for the United States Army, was responsible for spearheading the development of radio communications equipment for the military. Squier was a good fit for the mission. In the 1890s, while working toward his doctorate in electrical science at Johns Hopkins University, he experimented with remote detonation devices and radio-controlled artillery.

It's important to note that, before the United States entered World War I, communicating in the field via radio occurred through Morse code. The field radios that allowed soldiers on the ground to communicate with one another were cumbersome to transport, assemble, and knock down. It took up to three pack mules to transport one radio. The inconveniences of this form of communication in grimy battle situations inspired Squier to apply his knowledge of electrical engineering. He first improved radio communication by perfecting military radio tubes, creating standardized, rugged, and reliable alternatives to the finicky and fragile tubes available at the time. Squier also collaborated with established companies in the communication industry.

Next, he founded a research and development laboratory at Camp Alfred Vail in New Jersey, which became an intensive training site for signal troops. Radio intelligence operators at Vail not only learned to quickly intercept messages (in English and foreign languages). They also worked with the analog technology of the carrier pigeon. His staff of radio operators and engineers grew into the hundreds. It was out of this laboratory that the revolutionary Western Electric SCR-67 and 68 were crafted.[3] These radios were celebrated for their ability to transmit the human voice—a much more immediate way to receive information, rather than dots and dashes of Morse code. The SCR series was also designed for the reception of messages relayed between aircraft and ground radio operators. The quality of these transmissions was lacking; airplanes were loud and drafty, making them a less than ideal environment for clear communication. The SCR-67 (for ground use) and SCR-68 (airborne) radio units made the channels clear again. These radiotelephones not only resulted in a freer exchange of information between pilots and ground operators, they also allowed controllers to direct precise formation flying and acrobatics, and to guide airborne gunfire.

Squier's work with the SCR radiophone wasn't his only claim to fame. In 1909, he also received patents related to multiplexing telephony, a concept that is integral to the modern communication age.[4] His particular variety of multiplexing allowed multiple voice messages to be sent over a single wire. In today's terms, this concept is integral to networked computers and the flow of data that rampages through the Internet.

There is also the curious concept of "Tree Telephony." By hooking antenna wires into a tree—preferably oak, and in full foliage—Squier realized he could listen to broadcasts being issued from Nauen, the powerful German wireless station. Listening in to the wood via antenna leads, which ran from tree to transceiver to headphones, Squier bypassed the clicks of wood-eating insects to capture a different beat: rogue German radio signals, tapping out coded information from within the trees.

When the war ended, Squier's life as an inventor continued. In 1922, he filed a patent for "Electrical Signaling," which was the basis for his idea that, using existing electrical lines, business owners could receive piped-in music in their stores.[5] For a subscription fee, endless music would fill their shops through a wooden box provided by Squier. Intrigued with the made-up word "Kodak," he combined "music" with "Kodak"—and Muzak was born.[6]

Muzak is something we now equate with the sappy, soothing melodies heard in public spaces, like offices, department stores, and grocery stores. Muzak recedes into the background, creating a mind-numbing palette of sound. It's polite, it doesn't intrude. The sounds we tend to associate with Muzak are the result of psychological research—an effect called "stimulus progression."[7] Easy-listening tones build up in rhythm and intensity over the course of 15 minutes—the general lingering time of a shopper, or the time it takes a worker to perform an average task. Over time, Muzak was presented as a form of time and labor management, taking cues from Frederick Winslow Taylor's 1911 classic monograph *The Principles of Scientific Management*. This phase of Muzak, it should be noted, occurred well after Squier sold his rights to the invention.

A hidden trace of Squier is found, inexplicably, in a quilt. In 1891, the Dryden Ladies Library Association held a fundraising campaign to build a new Methodist church. They did so by selling quilt squares to prominent community members who paid to have their name on the quilt, and then paid again for tickets to the raffle to win it back. On the upper left-hand side of the Dryden Community Pictorial Quilt, the name George Owen Squier appears encased within botanical parentheses, along with the rest of his family. Each square of the quilt is embroidered in fine detail, depicting farm animals, furniture, tools, horseshoes, and flowers. One can't help but wonder if any of the embroidered Dryden farmsteads and businesses survived the heyday of Muzak as aural witnesses to its saccharine tones.

Squier was often featured in radio-culture magazines. The October 1922 issue of *Popular Radio* featured his image, along with the following caption: "Radio will bring to the people of this country the intellectual background which heretofore only the rich could afford. Yet the work of the radio engineer as an educator has only just begun. Soon we will be measuring culture by watts." The wooden box that accompanied the original Muzak subscription has essentially become our computer, one looped into digital music subscription services with readymade "chill out" playlists. Squier's "wired radio" has become truly wireless.

Notes

1. "The New Edison: The Phonograph with a Soul," *The Paris Morning News* (Paris, Texas, 13 February 1921), 12.
2. Ibid, 12.
3. Beauchamp, Ken. *History of Telegraphy* (London: The Institution of Electrical Engineers, 2001), 361.
4. US Patent #980357.
5. US Patent #1641608 A.
6. Lanza, Joseph. *Elevator Music: A Surreal History of Muzak, Easy-Listening, and Other Moodsong* (New York: St. Martin's Press, 1994), 30.
7. Lanza, 48.

RADIO CADIZ
Up The Hatch With The Piccards

An image taken from lofty heights, long before sunrise. Below, something that could be real, or could be a stage set—a hemisphere of something as soft and white as a sheet is rising from the ground. Blurred lights sketch out the boundaries of the tilted horizon. Car headlights snake to and from the makeshift launch pad, revealing access points to the site. This could be a clandestine image of a secret government project involving recovered alien spacecraft. Or it could be something less sinister. What is documented in this image is surprisingly (and accurately) connected to extraterrestrial explorations, albeit of the human-powered variety.

The image was taken on October 23, 1934 from a dirigible mooring mast overlooking Ford Airport field. The white object rising from the Earth is a partially inflated 175-foot balloon, growing swollen with hydrogen. Not visible is a lightweight titanium gondola awaiting its silent launch into the early morning sky, powered by grace of the (highly explosive) balloon.

The crowd of visitors and members of the press that gathered on the ground below are swallowed up by height and the dark grain of the photograph, though the lines of their parked cars are visible. Other views of this event, brought down to Earth, show school-aged children confusedly looking upward in all directions, as though the sky could answer why they've been hauled out of bed at 4:15am. In another staged media photograph, two angry-looking children plucked from the crowd, along with a few smiling adults, clearly of importance, are standing in front of a metal sphere with an open hatch. The sphere looks like contraptions daredevils use to launch themselves over Niagara Falls.

The hubbub captured in this series of images documents the Piccard balloon ascension, which rose nearly 11 miles into the stratosphere. A series of photographs taken in the daylight show cottony sheets of clouds and the view upward, into the sewn patchwork of the balloon as it ascends to altitudes of 9,000, 15,000, 16,000, and finally 57,579 feet, at full inflation.

Jeannette Piccard, who manned the gondola below the hydrogen-filled balloon, was a "street wise" lady with impressive credentials. She was a graduate of the University of Chicago's organic chemistry program and a fearless aeronaut. She was the first woman to be licensed as a balloon pilot, the first US woman to enter the stratosphere (and, technically speaking, space).

Jean Piccard, husband to Jeannette, was also crammed into the gondola, serving as chief scientific observer. Jean dealt with sensitive scientific instruments throughout the flight, gathering data on the nature of cosmic radiation. The flight's overall inspiration, beyond any notion of record-breaking, was born out of an initiative by the physicist Arthur Holly Compton, who hoped to conduct a worldwide survey to measure the velocity and intensities of cosmic rays.[1] The Piccards also brought along their pet turtle, Fleur de Lys, who became another record-setter as the first reptile in space.[2] The role of the turtle is undocumented, and it does not seem to appear in any photographs. Presumably, Fleur de Lys was brought along to study the effects of altitude on animals, over a decade before fruit flies and the rhesus monkey, Albert II, were launched from V-2 rockets at White Sands Proving Ground in the late 1940s.

On the day of the Piccards' flight, a record for high-altitude radio communication was set, as well. A shortwave radio had been built by William Duckwitz (with the help of William Gassett) to keep the lines of communication open during the Piccards' ascent. Duckwitz's design was based on plans taken from an article in *QST* magazine; its exposed knobs, wires, and tubes are typical of the DIY ethos and pure functionality found in amateur radio at that time.[3] Duckwitz's call sign, W8CJT, is burned into the wooden base of the radio's antenna as a point of pride—branded as a player in the hydrogen-fueled fervor of the Piccards' journey.

Radio equipment was installed in the gondola, and Duckwitz's homemade transceiver found a home in the backseat of a "radio car"—a standard Ford V-8—that trailed the flight on the roads below. Originally, it was calculated that the balloon would drift toward Cleveland, so the radio car headed to the city airport and attempted contact every half hour. But the day was overcast, and when the balloon was blown off-course, Duckwitz's radio didn't act as planned. As Piccard recalled in 1934, "Several times it was thought that the voice of Mrs. Piccard could be heard, but no useful information was forthcoming."[4] When an airplane managed to track the balloon, which, by this point, was beginning its descent, an estimate of where it was headed was radioed to Duckwitz. The V-8's engine made it possible for the radio car to blast down the country roads at speeds of 75mph. The caravan of newspaper reporters following behind couldn't hope to match this pace. Finally, near New Philadelphia, Ohio, "a signal was received from the Piccards which was quite loud but not very intelligible and the conclusion was reached that the cosmic ray apparatus must be causing Mrs. Piccard considerable trouble in operating the radio."[5] The Piccards seemed to have their hands full with other equipment while preparing for a bumpy landing.

While the Piccards were quickly shoveling as much ballast out of the gondola as possible, in order to stop their alarmingly fast descent toward the roof of a farmhouse, the radio car was speeding along the highway, heading for the vicinity where it was believed they would land. Piccard detailed the following: "As the radio car came to a stop the button controlling the siren, which had been installed so as to make better time through traffic, was accidentally touched and this added to the screech of brakes and ten men climbing hurriedly out of the cars caused the bystanders to believe that a holdup or a gangster killing was to be enacted before their very eyes."[6] This sonic blast from the radio car was too much for the locals, who were feeling justifiably jumpy after the events of the previous day. Just 50 miles north, a lethal shootout between the FBI and "Pretty Boy" Floyd had ended the bank robber's four-year fugitive run.

The Piccards "safely" crash-landed with only minor injuries at 2:40pm. They had travelled approximately 400 miles from their launching point, rolling to a stop in a forest outside of Cadiz, Ohio. With an official launch time of 6:57am, they had been in the air for almost a full eight-hour work day. One image of the aftermath of the landing shows balloon guy lines and rigging tangled up in the trees, as well as curious onlookers and a shaken-up Jeannette standing with her back to the camera, perhaps unready to face it. Another image, this time staged for the press, shows the couple posed in a happy embrace with their heads poking out of the gondola's hatch.

In 1986, a mashed-up legacy of the high-flying and deep-sea seeking Piccard family (August, Jean's brother, was also a famous explorer) would be honored by *Star Trek* creator Gene Roddenberry. The name Captain Jean-Luc Picard, fictional leader of "going boldly where no man has gone before," is no coincidence. Jeannette Piccard once said, "When you fly a balloon, you don't file a flight plan; you go where the wind goes. You feel like part of the air. You almost feel like part of eternity, and you just float along."[7]

Notes

1. "Strange Instrument Built to Solve Mystery of Cosmic Rays," *Popular Science* (New York: Popular Science Pub. Co., April 1932), 60.
2. Crouch, Tom D. *The Eagle Aloft: Two Centuries of the Balloon in America* (Washington, DC: Smithsonian Institution Press, 1983), 629.
3. Reminiscence, "Written Reminiscence of the Piccard Stratosphere Balloon Flight of October 23, 1934," ARCH.13, Box 18, Folder: Piccard Stratosphere Balloon Flight, n.d., Benson Ford Research Center, The Henry Ford, Dearborn, Michigan.
4. Ibid, 3.
5. Ibid, 4.
6. Ibid, 6.
7. Sorenson, Paul. "Looking Back..." *Aerospace Engineering and Mechanics Update* (Minneapolis: University of Minnesota Institute of Technology, 1998–99), n.p.

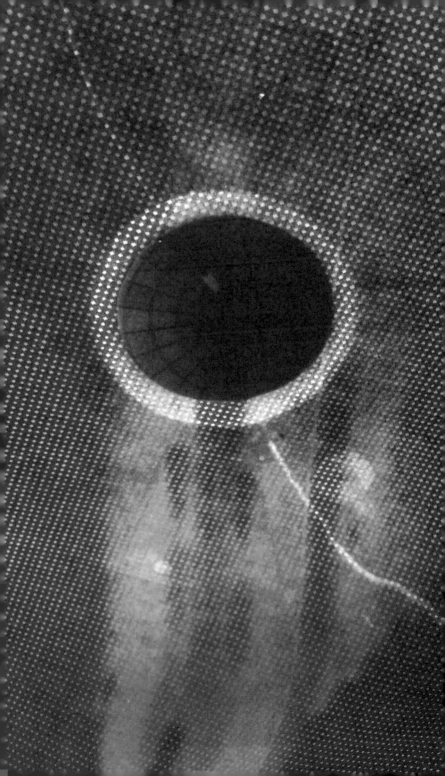

III. FREQUENCIES

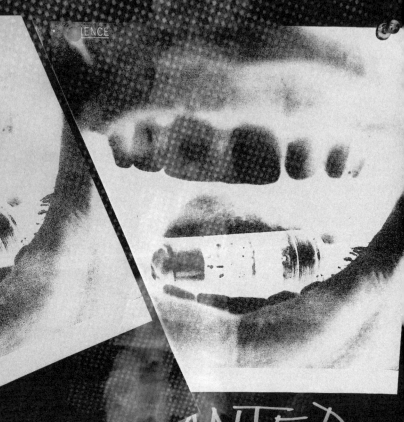

TUNE IN TO THE TUMMY
Vladimir Zworykin's Radio Pill

> "The time is coming when the doctor won't stop to ask you where it hurts. He'll just tune in on the radio and listen to what it says."
> — *New Scientist,* 18 April 1957

What if you could eat a radio station? A battery, transistor, condenser, coil, oscillator, and diaphragm—all crammed into the space of a capsule about an inch long and a half-inch in diameter. These components make up the "world's smallest FM radio broadcast station."[1] Developed in 1959 by Vladimir Zworykin, in conjunction with RCA, the Veteran's Administration Hospital, and the Rockefeller Institute, this medical grade "endo-sonde" was designed to sail through the thunderstorm of the stomach, and crawl through 10-odd meters of bio-sludge in the human intestine.

Ten thousand dollars for 15 hours of battery life is a tough pill to swallow, but that is exactly what Dr. John Farrar did. Farrar was the first test patient to ingest the prototype Radio Pill, as Zworykin watched. Before the capsule entered his gullet, he was sure to tie a paranoid string around it, in case an emergency retrieval became necessary. Exit through the entrance. In later iterations, if the Radio Pill stalled, a more level-headed approach was used: x-rays revealed its location, and magnets were applied to the skin to steer it back to the proper path.

The pill is cylindrical, with one end covered by a thin rubber membrane that vibrates to the beat of gaseous pressure waves exerted by intestinal fluctuations. These vibrations make their way to a diaphragm, then to an electric coil, then the oscillator. The oscillator, acting on information several steps removed, broadcasts a continuous radio signal to an antenna outside of the body.[2] The Radio Pill contains a resonant cavity structure that works to amplify sound—a structure not unlike that of "The Thing," an espionage listening device embedded in the carved wooden seal of the US Embassy in Moscow. The Thing was built (under proto-KGB duress) by Léon Theremin, the Russian inventor of one of the first electric instruments, now commonly referred to as the Theremin. The Thing was gifted to US Ambassador W. Averell Harriman as a "friendly gesture" in 1945.[3] Ironically, the foundations of The Thing were developed at Zworykin's own home-base of RCA in 1941.

In a 1961 Pathé medical demonstration film, a doctor examines a female patient and unwittingly plays her body as if it was a musical instrument, accompanied by sounds reminiscent of Theremin's own spacey instrument. The doctor applies pressure to her gut to change the pitch and tone of her amplified body, without ever actually touching the device transmitting within her.

Over the course of the Radio Pill's two-day journey within the body, an antenna-wielding doctor may tune in to the signal, which is bounced into a device much like an electrocardiogram, drawing waving lines across a piece of paper, or onto the screen of an oscilloscope. This addition of a visual record makes the body more like a television station than a radio station, as the pill transmits its data onto the oscilloscope's CRT screen.

(Zworykin's legacy as an electronics pioneer heightened with his invention of the iconoscope, which was the first practical vacuum tube designed for use in electronic television video cameras.) Despite all of these successful demonstrations of transmissions swallowed down with a sip of water, the pills were actually entrenched in the signals of war. As John Lear wrote in *New Scientist* in 1957, "The power for transmission of the signal is amplified by a transistor hooked up to a storage battery which was originated in WWII to set off anti-aircraft shells when they neared invading planes."[4]

The Radio Pill also used telemetry—technology-assisted exploration undertaken in dangerous or inaccessible landscapes— to explore the uncharted areas of the digestive tract, along with the deep and mysterious terrain of the right side of the colon. And somewhere in the bowels of institutions like the Henry Ford and the Hagley Museum, versions of Zworykin's Radio Pill have been swallowed up, too, locked within the tangled guts of object history.

Notes

1. "FM Signals Sent as 'Radio Pill' Passes through Body," *Electrical Engineering.* 76.6 (New York: American Institute of Electrical Engineers, 1957), 551.
2. Lear, John. "Now, the Broadcasting Pill," *New Scientist* (London: New Science Publications, 18 April 1957), 20.
3. Glinsky, Albert. *Theremin: Ether Music and Espionage* (Urbana: University of Illinois Press, 2000), 260.
4. Lear, 20.

CHARLES APGAR & SORDID FREQUENCIES

In 1911, the Stollwerck Brothers company purchased a 79-acre plot of land in Sayville, New York, reportedly to build a candy factory. Locals dreaming of confectionary employment, however, soon discovered that concrete pads were being poured to serve as anchors for a wireless radio station instead. The radio tower would reach heights of 477 feet, making it the most powerful communications hub in the area. In 1912, about 150 miles south of Sayville in Tuckerton, New Jersey, ships began unloading sections of an even bigger radio tower onto dry land. It was reportedly erected quietly, with no fanfare. Its construction was nearly complete before the US government realized it had started. How someone could *quietly* build two of the most powerful communications towers in the country—Sayville being an 820-foot behemoth only rivaled in height by the Eiffel Tower at that time—seems outrageous in retrospect.

The radio towers were soon discovered to be a ruse, and the chocolate factory had been a front all along. The owner of the stations, the Atlantic Communication Company, was the American subsidiary of the German telecommunications firm Telefunken. Some of Telefunken's principal staff had even been high-ranking members of the German army. The red flags of suspicion began to wave a little brighter.

In August 1914, the US government seized the Tuckerton station, claiming a lack of proper operating licenses. By the following spring, the Sayville tower—still under German ownership—increased the power of its transmitters to make up for Tuckerton's loss, and began skipping Morse code radio messages like stones above the ocean, all the way to Nauen, Germany. The lengths and bounds of this type of power in wireless communication, void of a relay station, was a feat previously unheard of, and proved to be a more powerful communication method than the German telegraph cable that the British had dredged up from the bottom of the sea, severing it in two. President Woodrow Wilson visited the Sayville tower soon after it opened, sending Kaiser Wilhelm II a strained-through-the-teeth "Happy 55th Birthday" greeting in January. In May, the British luxury liner *Lusitania*, which secretly carried munitions in support of British troops, was torpedoed by German submarines. This was the final straw, causing mass distrust of Germany's true intentions and a pledge of US commitment to the war effort. Posters depicting graphic interpretations of the lives lost on the *Lusitania* served as rallying cries for enlistment.

Three months later, with the Declaration of Neutrality in place, the American government was granted the right of refusal in issuing radio operating licenses to foreigners, and to prohibit citizens of "belligerent nations" from owning and operating wireless stations on "neutral" American soil. Just before the bill was filed, the US Navy grew wise and preemptively seized the Sayville tower. The seizures of Tuckerton, and later Sayville, mark some of the first "hostile actions" taken against Germany by the United States in WWI. Curiously, they allowed several Telefunken employees to remain in the US until 1917—an act of goodwill that later proved to be a mistake. Officials in the Navy were meant to monitor and censor Sayville's outward flow of communications, but secret workarounds were found, causing leaks in what was perceived to be a tightly run ship. Several messages were rejected by censors on the grounds that "they were too obviously not what they pretended to be."[1] What the self-negating messages actually *were* was yet to be known, though the content was far from hollow. Speculation at the time, and even now, connects the conspiratorial transmissions coming from Sayville and Tuckerton as the source of intelligence that led to the *Lusitania*'s loss, sparked by two foreboding words: "Get Lucy."

Charles Apgar was a salesman and broker by day and, by night, an amateur radio operator at the Westfield, NJ station W2MN. In the early days of radio, before voices sang over the wires, Apgar listened in to radio traffic in the form of dots and dashes, which he would convert into legible information: weather reports, shipping chatter, time stamps. In 1913, Apgar became a fledgling inventor when he jerry-rigged an Edison wax-cylinder device to record these radio transmissions. Apgar's recordings are believed to be the first to permanently commit wireless telegraph signals to physical playback media. His desire to record the late-night signals and static sent out of Sayville was born out of a gut feeling of suspicion. The Chief of the US Secret Service, William Flynn (known as "the anarchist chaser") was quietly apprehensive about the station too. When he heard about Apgar's recording contraption, he promptly asked him to "get busy."[2] Perhaps it was innocent, or perhaps this was tongue-in-cheek, but "get busy" reads as a close vengeful cousin to the doomed message "Get Lucy."

Like a modern DJ crossfading records on a double turntable system, Apgar set up two phonograph machines and switched an Ampliphone receiver back and forth, which allowed him to change out cylinders as they filled up—all without missing a beat. In an article titled "How I Cornered Sayville," Apgar writes, "That not a single message was missed and hundreds were recorded is evidence that every instrument and device of my home-made set did its duty fully and promptly; which is to me, of course, very gratifying."³ The same article cites a Dr. Frank of the Atlantic Communication Company, who seemed threatened by Apgar's airwave prowess: "The statement that Mr. Apgar can record messages sent out by wireless on a phonographic cylinder is hardly worth discussing. That is physically impossible. [...] If Mr. Apgar has accomplished it he should get his idea patented and perhaps we will buy it."⁴

But Apgar wasn't lying. For 14 nights straight, between 11pm and 2am, Apgar built up his stockpile of "canned" wax recordings, handing them off to Secret Service encryption specialists. Once Sayville sent out its nightly test signal, Apgar had to work quickly to synchronize his own tuning: "There were several circuits to be looked after, their various condensers, inductances, batteries—everything, had to be practically in perfect harmony the instant Sayville began sending."⁵

It turned out those vague suspicions were not unfounded: Overtly commercial messages were being run through with a cipher code filter. The curious spaces, repetitions, and extra letters sprinkled into censor-approved messages were providing intelligence to the German U-boats, which circled like sharks in deep-sea waters. The messages travelled far, point-to-point—from Sayville to Nauen, and then relayed back out to the submarines, which had now grown wise to the location of neutral ships. The human ear alone couldn't be blamed for missing the security leaks. Apgar's phonograph recordings helped to amplify the issues: embedded acrostics in the transmissions; "time spacing"; words slipped between "mistakes"; repetitive spaces and letters of suspicious variations; and an overemphasis on "message checks," like a fax confirmation. This added noise was not innocent. The secrets hidden within the spaces were passing right under the noses of US censors. The Secret Service thanked

Apgar for his service, then promptly put legislation into place to ban all amateur radio activity for the remainder of the war.

This wireless radio equipment was used to reveal the vulnerabilities of the American airwaves in the WWI era, and played a role in publicly exposing this breach. The two towers at Tuckerton and Sayville, like Apgar's recordings and transcriptions, are now lost, but his radio equipment has been preserved, as have its effects. What survives is a wooden ghost box with dials, and a stacked metal guillotine trap to filter and tamp down ethereal radio waves for human ears. Peeking under the metaphorical lids of these objects reveals a sordid story of sound, unbound. A curious world of ciphers, codes, proto-cybersecurity, and the value of amateur ingenuity.

Notes

1. "The Quenching of Sayville: The Close of a Sordid Story," *The Wireless World* 3, no. 32 (November 1915): 515.
2. Apgar, Charles E., "How I Cornered Sayville," *The Wireless World* 3, no. 32 (November 1915): 519.
3. Apgar, 519.
4. "The Quenching of Sayville," 517.
5. Apgar, 519.

CALLING, HAUNTING

When K. was in fourth grade, her cousin told her how to make the phone ring. They dialed the test-tone number #519-999-627-5713#0000. The phone rang, and it rang longer and louder than it should have. The sound seemed ominous, so she slammed the phone down. It kept ringing and, just as her fear rose to a pitch, the noise stopped. Later, when her mother was home, K. dialed the numbers again. Her mother didn't know about the test-tone ring, and was convinced it was her dead aunt calling, harnessing the phone lines to make her presence known. K. perpetuated the haunting for a few days, making the telephone ring whenever her mother was around. She soon grew bored and stopped.

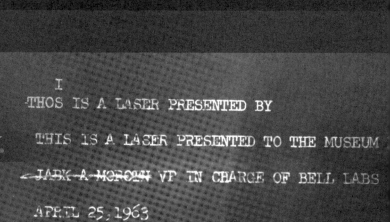

THOS IS A LASER PRESENTED BY

THIS IS A LASER PRESENTED TO THE MUSEUM

~~JACK A. MOROTH~~ VP IN CHARGE OF BELL LABS

APRIL 25, 1963

JACK A. MORTON

ALL TOGETHER NOW, SPEAKING ON A BEAM OF LIGHT
An Impure History of the Maser and Laser

Lazer, anesthesia, maze ya,
Ways to blaze your brain and train ya
—Public Enemy, "Welcome To The Terrordome"

I. *Assembly*

On December 12, 1960, Ali Javan and William Bennett, two researchers at Bell Telephone Laboratories in Murray Hill, New Jersey, tested the first helium-neon gas "MASER" (Microwave Amplification by Stimulated Emission of Radiation). When technical developments made it more logical to replace the word "microwave" with "light," the more familiar term "laser" emerged. The difference between the laser and the maser is in one little letter. As a device that began its life working in the microwave spectrum, a short shift transferred its effects into the optical areas of ultraviolet and infrared. The same letter 'L' suggested this was a device that could generate light in a controlled way.[1]

A clear-cut lineage of masers and lasers—a lineage as clear as laser-cut glass—is rather chaotic. Their histories and functions are intertwined, a confusion that belies the precisely ordered physics of these devices. At the risk of irking my fellow technology historians, the invention and function story is not the point of this essay. I'd prefer to present a series of sub-histories, material archaeologies, creative hijackings, and conspiracies surrounding the maser and laser.

The core history of the maser and laser is contingent upon a few essential moments. The maser began as a university-funded theory on paper, shucking the fantastical trope of "birth by sparking science fiction" in a mad-scientist laboratory. Two papers[2] written in 1954 and 1955 by Charles Hard Townes, James P. Gordon, and H.J. Zeiger outlined the team's explorations with the microwave-emitting maser. In 1953, these same three physicists, led by Townes, built an ammonia maser the size of a small refrigerator at the Columbia Radiation Laboratory. Truthfully, maser was an invention without purpose; it was a "solution in search of a problem."[3] Townes established himself as a key figure in the history of maser and laser technology from this point onward. But all notions of there being a glowing line of light illuminating Townes's eyes, beaming with retinal pride at this early date should be scrapped, at least temporarily.

The emissions of the first pulsing ruby masers were demonstrated via their influence on other materials. A powerful flash of light left char marks on a target travelling at speeds so fast that its line of action was rendered invisible to the naked eye. Then there's the laser as portrayed in the classic heist movie—a basket of woven, colorful security beams that exist for the sole purpose of being bypassed by the controlled tumble of a masterful sneak who snatches a treasure illuminated by a perfect downward cone of light. The archetypical beauty of this kind of uncontained red beam-blast came later, once maser research moved into the optical range in 1960.

The first mental marriage of lasers and visible beams of light was once again theoretical, appearing in the essay "Infrared and Optical Masers,"[4] written by Townes and Arthur Schawlow and published in *Physical Review* in 1958. The essay essentially sparked what would become a global race to turn the technology into a reality. Their idea remained confined to the page for two full years, before the first physical light-emitting optical maser was finally announced to the public—made viable not by Townes, but by others. This is the point at which the maser shifted taxonomy and became the laser. The optical maser and laser became interchangeable words with identical meanings.

Townes and Schawlow were subsequently granted US patent #2929922, "Maser and maser communication systems." This signified a device where "light itself could be amplified by making the waves travel back and forth along a long thin column of amplifying substance that was contained between two very flat mirrors."[5] In this scenario, one mirror is polished to be highly reflective; the other, not so much. The beam of light feeds back upon itself, travelling through an "amplifying substance" (usually gas) between the mirrors. This is how feedback is introduced into the equation. The feedback excites the gas atoms into a higher state, and "stimulates" them by microwaves. The light bouncing between the mirrors that isn't travelling at perfectly right angles is allowed to escape. The light that remains isn't diluted, but is radiant, intense, and coherent. It is then stimulated into concentrated light or high-frequency radio waves. This, at its most basic physical element, is what creates the coherent beam of microwave radiation, or stimulated escapist energy. This is the base maser beam. In a highly simplified alternate analogy, if a stone is thrown with a strong left hook against the wall of a well, the clack and bang and "plop" when it hits the water at the bottom will resonate back up to whoever has ears dangling over the entry to the well. This is amplified sound observed from the top of an echo chamber. The excitement of the maser and laser is similar—a process in a contained, resonant space where feedback continuously loops back upon itself, until it gains enough momentum to become something else. In the well, the echo effect has shifted an action into new aural forms. In the laser, the production of "coherent light" is more intense than "normal" light. At least, for a non-scientist, this is an approximation of how the laser operates.

While Townes and Schawlow were working at Bell Labs, the theoretical aspects of the maser were also being explored independently in Russia by Nikolay Basov and Alexander Prokhorov at the Lebedev Physical Institute. Coinciding with the first Townes papers, the Russians' first paper was published in October 1954. Townes, Basov, and Prokhorov would share the Nobel Prize for Physics in 1964.[6]

The title of first practical laser was claimed by Theodore Maiman at Hughes Aircraft Company in May 1960. This closely coincided with Ali Javan and William Bennett's work at Bell Labs in December of the same year. Maiman's maser sent stuttering pulses of light through a ruby (pink, not "pigeon blood") into a xenon-gas-filled tube. The Bell Labs version produced steady, continuous light by discharging an electrical current through helium and neon gas.

Coinciding with the early years of maser research, the military provided sufficient funds for research and design projects that accelerated electronics. Electronics were at the heart of powering missiles, which could be used against the invisible threats of the Cold War. Ironically, optical devices were simultaneously considered as anti-ballistic devices, and laser and maser communications were somehow immune to the tabula rasa magnetic boom of nuclear weapons.[7] Federal military funds for electronics research were increased dramatically in the years of and following the Korean War (1950–1953)—a trend that continued until "the euphoria was over and a period of more sober research succeeded."[8]

II. *Chatter*

Optical masers produce concentrated, clean, and sharp beams of light—the type of light that could serve well as a noise-free information carrier. The microwave's short wavelengths produce higher frequencies and, in turn, these high frequencies are able to hold more information. Due to its narrowness, the frequency beam of the maser can be extremely fine-tuned. In this case, the term "talk sharp" becomes literal.

A 1963 educational film, *Principles of the Optical Maser,* claims that "a single maser beam might reasonably carry as much information as all the radio-communication channels now in existence." Or, a similar comparison of simultaneity: Every living human could pick up the receiver to hold a conversation with another person. The speculative scenarios of these maser-guided communiques imagine a super-loading of the network—a future when all able-eared (and mouthed) citizens can participate in a global phone-in, their conversations riding alongside one another, on beams of pure light that illuminate the grid.[9]

The potential for synchronized chatter with the maser is comparable to the excitement that followed the laying of the transatlantic telegraph cable in 1858. Soon, multiplexed conversations were tamped down into dots and dashes. Shapes of sounds, switch-back riding through the wires of barnacled hemp and steel tube lying on the Atlantic Ocean's seabed, stretching between Heart's Content, Newfoundland and Valencia, Ireland. This "wireless" method of transmission was as murky as the silt it laid in, but it was vital to connecting continents and cultures at the time. Submarine telegraph cables are riddled with valorous stories of wealthy technological dreamers forced into situations of seafaring pluck. As more cables were laid and connections made, wires began to wave around the ocean floor like mats of seaweed. They're still down there.

The year JFK was sworn into office, Javan and his crew were burning spots into their vision and swigging secret celebratory libations from beakers at the Murray Hill laboratories.[10] In December 1960, Javan took part in the first telephone conversation using lasers as carriers for human speech. The day after he and his team successfully created the helium-neon maser beam, Javan called the lab from home after a late start. "One of the team members answered and asked me to hold the line for a moment," Javan recalled in an oral history from 2012. "Then I heard a voice, somewhat quivering in transmission, telling me that it was the laser light speaking to me." The voice, transposed to laser light, was transmitted as a light beam shot across the lab Javan worked in, discharging into a light detector and broadcast as a voice signal into the telephone.[11]

Elsewhere, speech becomes fractured—spectral, just for a moment. It is dematerialized down into those submarine wires, a new layer in the substrata of the ocean or earth to be reassembled. Receiver to receiver. The lack of such a robust support grid hushed the voice of the maser before it ever had a chance to compete. Maser light was free of congestion. But it was sensitive to external interference when shot freely through the air, as discovered by a team of Bell Labs scientists when they beamed "a pulse of laser light twenty-five miles through the air from Crawford Hill in Holmdel to the Murray Hill labs."[12] It's easy to conjure ambitious language when imagining this event. Rather than the intense flash of ruby-powered light, a steady laser beam being "cannoned" or "shot" across those 25 miles is an easy mental exaggeration to make. Weather patterns and the bend of the Earth, ungrounded static and storms—all jockeying to splinter the purity of the signal. A resonant sheath of some kind, like the hemp and tar that protected the telegraph cable, was needed to contain the light.

Enter the age of fiber optics. By trapping free-roaming maser signals into tubes of fine glass, a new communications medium was born:

> A coherent waveform, on the model of radio waves, can function as a carrier, with modulations to it decipherable as signals. After experimenting with airborne transmissions (rain, fog, and air pollution degraded the signal unacceptably), experimenters in the US, UK, Russia, and Japan eventually converged on getting coherent light from lasers to travel through flexible glass pipes: fiber optics.[13]

This is the maser-laser beam contained—language broken into signal, knocking against once-molten silica doped with germanium and fused into hair-caliber fiberglass threads that are the purest of the pure. There are shades of the hip-hop artist and language visionary Rammellzee here—something Gothic Futurist about such a reduction. We might imagine pure speech travelling through glass as energy. Or as symbolic warfare, as though the words lobbed from an earpiece to a mouthpiece would travel like sentient balls of light, exploding upon their arrival as words bomb into the ears of listeners.

A standard plane of glass, turned on its edge, shows a boundary line with hues of blue or green. These are impurities and the condition that leads to a "bad connection." Fiber optic glass strands rely on uncontaminated materials; the strands were a new form of manufactured purity invented by the Corning, Incorporated glassworks. These strands would allow utter clarity of signals, as well as access points into the backbone of what would become the Internet and the delivery of mainlined digital desire. Rogue signals now flash through tubes of light, tugging at the glassine hair of the Rusalki[14] stuck in the drainpipe, trying to escape from the tangle and blast out into dark bodies of water.

III. *Purity*

Photographs of the helium-neon optical maser in operation make everything *but* the coherence of light fade away. The device is photographed in black and white, in the dark, to make us unquestionably aware of its overpowering inner light. The intensity of the maser beam is pronounced—a high-contrast addition of light in the darkness, swelling to create space for itself within a concentrated black absence. All trite comparisons of light conquering darkness aside, the maser is light on overdrive. Photographed in color, the helium-neon beam would glow red. The pamphlet to accompany the 16mm film *Principles of the Optical Maser* reflects this in the form of a high-key color exploration—purity of red against black.

In contrast to the rebellious laser, which is free of containment, other sources of light are produced and amplified specifically through conditions of restraint. The filament of the incandescent bulb is protected by its thin shell of lamp glass. The fluorescent tube creates a nearly imperceptible vibrating pitch that grows louder as its lifecycle ends—mercury and ionized argon and arcs and ballasts all working to entrap the buzz.

Ordinary light is "incoherent." Frequencies come from all directions and phases. A display by Bell Laboratories in the early 1960s characterizes these types of light sources as suitable for only "the crudest signaling purposes."[15] Marine lanterns and smoke signals and searchlights be damned. At the opposite end of the frequency of legibility, if ordinary light is "incoherent," a laser is given the clear-headed attribute of being "coherent." This is rooted in the thin beam of light, its single frequency and utter purity, producing "light waves exactly 'in step' with each other."[16] The light of the maser and laser both operate on an ordered logic. It's a characteristic that turns a *process* into an *object*—a thing to be controlled. There is no better quote to fall back on than Marshall McLuhan here: "The electric light is pure information. It is a medium without a message…"[17] But the optical maser resists this type of thinking. It is synthesized light, unintended for illumination, and is a carrier of all that McLuhan is negating.

IV. *Dazzle*

The differences between the laser and maser are distinctive in some areas, grey in others. John Pierce—Bell Labs engineer, researcher of communications satellites, and memorable maker of techno-quips—said, "The laser is to ordinary light, as a broadcast signal is to static."[18] He also said, "Funding artificial intelligence is real stupidity."[19] For communications purposes, the maser is to satellite and radar what the laser is to holography and automated sensing. The maser can penetrate living tissue or clouds without causing disturbance. But microwaves can also cause the sensation of searing flesh, as evidenced by the military's non-lethal directed-energy weapon, the Active Denial System, or the ability of the US Navy's Laser Weapon System (LaWS), which is mounted on destroyer ships, to target and "kill" the motors of smaller attacking boats. Beamed dead in the water, rather than smoke on the water. Laser light can, depending on how far its intensity is dialed, act as a guide or as a weapon. High-powered lasers can be violent, reductive. They can cut through wood and steel, and drill and weld automotive parts.

When dialing back the juice, the laser in its gentler forms is typically used to hang levelled picture frames. But the same red laser dot can turn deadly when it supplements a gunsight. The "laser dazzler" is used in military conflict to temporarily stun and confuse enemy vision—or, a consumer-grade laser pointer might be used in the same way. This occurred at a 1998 Kiss concert, when some joker shone his laser light in the eyes of drummer Peter Criss while the band was performing "Beth."[20] But the Catman didn't want to play; a pissed-off Paul Stanley stormed the mic and mouthed off to the crowd.

Shining a laser onto the body of a police officer or into the cockpit of an airplane is illegal, and has more deadly potential than getting told off by the Starchild. Although, case in point, amateur astronomers *do* sometimes use green beam lasers on moonless nights to sketch out constellations for bystanders.

In the 1950s, toy death-ray guns existed before the maser was ever fully understood by the public. By the late 1970s, there were toy signaling flashlights and luminous tubes meant to fuel fantasies of interstellar adventures and play-wars over the unclaimed territories of outer space. Ironically, directed-energy weapons were undergoing serious military study at the same time. But due to the expense of commercial laser diodes, these toys lacked *real* lasers—they only referenced the *idea* of the laser.

V. *LASER: ART: MEDIUM*

As the maser developed beyond its military roots, technologically driven artists sought ways to harness it to create immersive environment works. Laser light was explored as a medium—as an *object*—by artists in the mid-1960s and 1970s. Its physical nature was manifested as a presence that could modulate space, render signals visible, and reference purity, both in terms of its color palette and its coherence. In some ways, the wondrous effect of the laser—those plasmatic beams, working within highly controlled scenarios—harken back to the "pure spiritual forms" of glowing ectoplasmic productions conjured up in Spiritualist séance rooms. Creative frauds, for the most part, worked in dark chambers, calling up seemingly untamable forces. Belief and wonder (combined with light effects) were commonly exploited in the science demonstration, the séance room, and the art gallery.

The laser relies on the persistence of vision. Viewers experience its visual presence via the vibrational afterimage. But searching for visual documentation of these early projects reveals that the laser's afterimage has mostly faded. There is a misalignment between the level of excitement with the new medium of laser technology and the degree to which its deployment was visually documented. Only a small handful of images appear to exist of David Tudor and Lowell Cross's *VIDEO LASER II*, a laser system displayed at the 1970 Pepsi Pavilion in Osaka, Japan. These projects have receded into the background, as muddy as the exterior of the building shrouded in a misty "cloud sculpture."

As a new medium in the arsenal of technologically minded artists, the laser was soon linked to the vision-sound paradigm. To make sound a visual experience, Lowell Cross began experimenting with electronic music in the mid-1960s, after he had "connected an RF modulator to a television receiver to interpret his own music as well as the works of composers John Cage and David Tudor."[21] By 1970, Cross was collaborating with David Tudor at the Pepsi Pavilion in Osaka, where he and Carson Jefferies used Tudor's music to influence the patterning of the laser system they installed in the pavilion's darkened entry room, known as the Clam Room.

In 1967, Robert Whitman's *Straight Red Line* was installed at the Pace Gallery in New York. The following year, Keiji Usami created a "walk-in" laser show titled *Laser:Beam:Joint*. Why the abuse of all of these colons? And why the direct naming of materials? Most painters wouldn't think of titling a painting *Painting:Oil:Sailboat*. In any case, Usami, working as part of the E.A.T. collective (Experiments in Art and Technology), shot a red and blue laser beam through green plastic panels that depicted running men. The viewer, standing in the middle, experienced "a network of symbolic relationships." Two other members of E.A.T., John Anthes and Tracey Kinsel, created the ELLI (Electronic Laser Light Image) project at MoMA in New York City in 1967, which produced laser light images controlled by audio signals. Automatically generated signals were either produced by magnetic tape loops or by audiences who were invited to contribute to the work by pressing keys on a synthesizer keyboard. Complexity could be added to the image in this way, through the use of chords.

According to the MoMA catalog, ELLI was "a three-dimensional light image that responds to information given to it. The viewer can have a dialogue with image (extension of ELLI's soul) … the image has dialog with surroundings."[22] At Caltech in July of 1969, the physicist Elsa Garmire[23] worked in her downtime to create the "Cybernetic Moon Landing Celebration," an argon laser light "wall" to celebrate the US lunar landing. In 1969, the Cincinnati Art Museum hosted *Laser Light: A New Visual Art*, credited as the first major laser light exhibition in the United States. Within the exhibit, several artists explored the capacity of sound vibration to affect laser beam patterns, including James Rockwell's *Laser Beam: Synthesizer*, which showed "a changing oscillating pattern on a rear projection screen to the electronic sounds of a twelve harmonic wave synthesizer."[24]

Science centers gobbled up permanent laser demonstrations, and art galleries became temporary science centers. Viewers of these early laser art installations seemed to determine that no recording medium could capture the intensity of the experienced reality, and thus chose to document little. By the time art historians thought of lasers as anything beyond a misguided infatuation with new technology (or a penchant for schlock) beam aesthetics had been rendered mundane by the ubiquity of the technology.

Ivan Dryer, an aspiring filmmaker, was introduced to the laser when he visited Caltech in 1979 and witnessed Elsa Garmire's laser projections. Dryer is in fact quoted as saying that no film stock could do justice to the saturated sublimity of Garmire's laser-vision justice. He also came to another conclusion: "When the laser turned on, so did I."[25] With a bit of convincing, Dryer's personal ode to the laser—an hour-long live "Laserium" show—was displayed at the Griffith Observatory in Los Angeles in 1973. Visitors imbibed the substances of their choice in City Park before experiencing an hour of whiplashed necks as they took in prismatic explosions on the ceiling of the Samuel Oschin Planetarium. The "laserist" kept abstract time with the classic rock of Emerson, Lake & Palmer, Pink Floyd, and the Rolling Stones; the Moog-y envelope generations of Wendy Carlos; and the straight-up classicism of Johann Strauss.

By 1984, Pink Floyd's *Dark Side of the Moon* Laserium concert became a staple for teenagers and awkward first dates. The Laserium created an impermanent architecture, given temporary lease to play within a historic site. The domed ceiling of the Griffith Observatory, built in 1933, distorted into a "media temple" that Dryer referred to as "new entertainment" working in the "environmental" approach.[26] "What we're talking about now is the creation of new realities," he said, "indistinguishable from the old one we now share, at least in their verisimilitude." A 1978 issue of *Arts Magazine* claimed the Laserium contained "seeds of what will become the high, universally acclaimed visual art of the future." High, indeed.

VI. *Novelty*

While a presentation laser pointer in 1981 cost hundreds of dollars, cheaper diode production finally led to affordable lasers by the 1990s. The laser in its modern form is now miniaturized, diverse, and abundant. It's in the barcode scanner at the grocery check-out line. It scans Blu-ray discs and corrects vision via the Lasik procedure. It's at the heart of computer printers and holograms. It removes hair from unwelcome places.

The laser, now moving toward the top of the novelty pile, is also what powers the common laser cat toy. As the cost of laser production plummeted, a crucial moment in its ubiquity coincided with the 1993 patent "Method for Exercising a Cat."[27] The sole patent listed under the names of Kevin T. Amiss and Martin H. Abbott, the patent applicants note that "cats are not characteristically disposed towards voluntary aerobic exercise," and provide a means to change this:

> The coherent nature of a laser light beam results in a small intensely bright pattern of light clearly visible in normal day light or artificial night illumination, small enough relative to the paw of the cat to cause interest without posing a threat, and sharply defined over long enough distances (e.g., up to 150 feet) to provoke a full workout with long sprints for the pet. Ideally the bright pattern of light is directed along the floor, steps or wall at speeds sufficient to exert and entertain the cat but not so discouragingly fast as to dissuade against the chase, i.e., typically in the general range of 5 to 25 feet per second.[28]

A series of adventurous iterations of Amiss and Abbott's brainchild soon followed, including a patent for a "Method and apparatus for automatically exercising a curious animal."[29] With direct reference to the types of laser pointers that 1990s-era businessmen used to circle eye-boggling complexities in their PowerPoint presentations, and that art historians used to highlight browned-out Kodachrome slides during lectures, the laser was denigrated to the position of stirring lazy housecats into action.

VII. *Darkside*

Back when scientists were starry-eyed about the applications of the maser—before the SETI Institute was formed,[30] before man had landed on the moon—Townes, now working with the scientist R.N. Schwartz, was already considering its far-flung possibilities. The maser could theoretically amplify the fields of radio and infrared astronomy, with "the possibility of interstellar communication by radio-waves in the microwave region ... to search for signals from intelligent beings on planets associated with nearby stars."[31] With the assumption that some advanced society on a brother planet would have a sister earth ray, an agreed-upon signaling language—and the naked human eye—would be the only tools necessary to hold a conversation light-years apart, powered by maser beams tumbling from one planet to the next. The earthbound end of the maser chain would be more powerful if put into practice on "a very high altitude balloon, a space platform, or natural Moon—a possibility that may not have been seriously considered a few years ago, but should be more acceptable today."[32] On May 12, 1960—four days before Maiman laid claim as inventor of the first working laser—the following was published in *New Scientist*:

> Will it be possible, one day, to make a beam of light of such intensity and so nearly parallel that one could project a visible spot of light on to the dark limb of the Moon? ... [E]ven if it did [exist] the construction of a lunar sign-writer would be daunting indeed.[33]

At the cusp of a moment, popular science was angling for the most radiant future possible for the maser in the form of "moonwriting." By 2015, however, these shots in the dark mirrored an image of Planet Earth, talking itself into a feedback loop, as we witnessed the Curiosity Rover on Mars make track marks in the sand. As Curiosity roamed Mars, searching for signs of ancient habitation, it zapped rocks to discover chemical compositions with its infrared laser ChemCam, and its maser-boosted communication systems beamed information back to NASA.

The Spirit rover, in one of its last hurrahs, harnessed the power of humanoid pattern recognition and immaturity. While running in land exploration mode and banking a sharp turn, its six-wheeled advantage left a permanent joke in the sand: a giant Red Planet penis.[34] The Face on Mars, popular in 1980s tabloid newspapers, couldn't compete with the modern trolling power of the Internet, and The Face melted back into the mesa it always apparently was.

Jack Morton, vice president of electronic technology at Bell Labs, personally presented a rare working model of the helium-neon gas maser to the Henry Ford Museum in 1963 at a meeting of the Institute of Electrical and Electronics Engineers. Audience members were treated to a selection of scientific films created by Bell Labs and a lecture by Morton titled "Cracking the Numbers Barrier in Electronics." When the maser was presented at the museum's theater by Morton to then-curator of technology, Frank Davis, it was done with the belief that the institution would be a fitting repository for this device, and the cataloger of its history.

The first transistors—the pervasive nervous system that would power everyday life in the electronic age—were transformed into viable products under Morton's leadership at Bell Labs in the late 1940s and early 1950s. In an odd conspiratorial twist, the transistor project has been claimed by some to be an outcome of "recovered and replicated alien technology" from the purported 1947 UFO crash at Roswell, New Mexico.[35] Some naysayers claim the efforts being undertaken at Bell Laboratories were beyond this earth—beyond human intellectual capability—and that shrinking the power of the vacuum tube into something the size of a pencil eraser should be credited to the grey men with bug eyes.

Morton, of course, purportedly played a direct hand in this cover-up. How else could he have written the report on transistors requested by his temperamental boss, Mervin Kelly, in a week's time? In 1971, in the middle of the night, Morton's Volvo was transformed into an inferno outside of New Jersey's Neshanic Inn. Once the smoke cleared, the gasoline-soaked husk of his body was found inside.[36] Some believed it to be part of a murder-for-hire conspiracy. Jack's alcohol-fueled sloppiness made him a time bomb who could unleash "the truth" at any moment, and he was silenced accordingly.

A fragment of text typed on the maser's accession record catalog is filled with uncertainty: "Tho^i^s is a laser presented by / this is a laser presented to the museum bh / ~~jabk a morotn~~ VP in charge of Bell Labs / April 25, 1963." The name "Jack A. Morton," spelled correctly, was handwritten in, to clarify. Included with the maser was a set of instructions for its operation. Crystal maser clear. General Instruction #1? "Never look directly into the maser beam."[37]

Notes

1. The Bell Laboratories film *Principles of the Optical Maser* defines the technology as such: "An optical maser is an optical oscillator, in many ways similar to a radio oscillator, except that it gives out light waves instead of radio waves. ...Radio and light are both electromagnetic waves."
2. These articles are: "Molecular Microwave Oscillator and New Hyperfine Structure in the Microwave Spectrum of NH3," J.P. Gordon, H.J. Zeiger, and C.H. Townes, *Physical Review* 95 (1 July 1954), 282; "The Maser—New Type of Microwave Amplifier, Frequency Standard, and Spectrometer," J.P. Gordon, H.J. Zeiger, and C.H. Townes, *Physical Review* 99 (15 August 1955), 1264.

3. The quote is often cited, but its origin is apocryphal. It is often attributed to either Irnee D'Haenens or Theodore Maiman. See Jeff Hecht, *Beam: The Race to Make the Laser* (New York: Oxford University Press, 2005), 9.
4. The pair had been thinking along parallel lines since the late 1940s. Schawlow worked with Townes while on a fellowship at Columbia. Together they would co-author the book *Microwave Spectroscopy*. Schawlow left the university, but was later reunited with Townes, in the mid-1950s, when they began maser research in their spare time.
5. *Laser Light; a New Visual Art: Organized by the Cincinnati Art Museum and the Laser Laboratory of the University of Cincinnati Medical Center, Cincinnati Art Museum, November 12 Through December 14, 1969* (Cincinnati Art Museum, 1969), 7–8.
6. Albert Einstein is the source of maser history. In his 1917 paper "On the Quantum Theory of Radiation," he lays down hard facts about stimulated emission. But he was a busy man, and Einstein left his theories as speculation on paper for others to make viable.
7. Cubitt, Sean, *The Practice of Light: A Genealogy of Visual Technologies from Prints to Pixels* (Cambridge, Mass: MIT Press, 2014), 224.
8. Bromberg, Joan L., *The Laser in America, 1950–1970* (Cambridge, Mass: MIT Press, 1991), 12.
9. Alexander Graham Bell, for his part, invented the photophone in 1880, 73 years before the maser. He threw his "speech on a beam of light" hat in the ring when he discovered that reflected sunlight could be modulated with sound vibration. See US Patent #235496A.
10. Bell Labs had an official ban on liquor in their employees' work areas, but optics specialist Don Herriott bucked the rule with a celebratory drink. Ed Ballik, a technician, supplied the libations. See Jeff Hecht, "History of Gas Lasers," *OPN Optics & Photonics News* (The Optical Society, January 2010), 17. According to an entry on Ali Javan's personal website, a secretary pushing a mail cart distributed notices to staff the following day: "At Bell Telephone Labs, we want to reiterate that there shall not be any alcohol beverages served or consumed in the premises EXCEPT they are over 100 yrs old."
11. For an extensive personal oral history written by Ali Javan see "The Historic 12 December Event: The Year 1960," *The History of the Laser* (2012). <www.alijavan.mit.edu/Original/Historic/12December1960event.html>. Accessed 9 October 2017.

12. Garrett, CBG, "The Optical Maser." *Electrical Engineering.* 80.4 (1961), 248–251.
13. Cubitt, 224.
14. The Rusalka is a mythological female water deity found in Russian and Scandinavian folklore. Loosely translated, the Rusalka shares many qualities with the mermaid.
15. Explanatory image supplied by Bell Laboratories, found in an accession file at the Henry Ford Museum, Dearborn, MI: 63.173.1, Laser (optical instrument), "Bell Laboratories Helium-Neon Gaseous Optical Maser, 1963."
16. L*aser Light: A New Visual Art,* 7.
17. McLuhan, Marshall, *Understanding Media: The Extensions of Man* (New York: McGraw-Hill, 1964), 15.
18. Gertner, Jon, *The Idea Factory: Bell Labs and the Great Age of American Innovation* (New York: Penguin Press, 2012), 255.
19. Quote gathered from the liner notes of the vinyl re-release of music programmed by John Robinson Pierce on an IBM 7090 computer, *John Robinson Pierce: Music From Mathematics, Volume Two* (Finders Keepers Records / Cacophonic: CACK4504, 2013).
20. "Laser Pointer Irks Kiss," *Beaver County Times* (Beaver County: PA: McClatchy-Tribune Information Services, November 24, 1998), n.p.
21. Daukantas, Patricia, "A Short History of Laser Shows," *OPN Optics & Photonics News* (The Optical Society, May 2010), 44.
22. Hultén, Karl Gunnar Pontus, *The Machine as seen at the End of the Mechanical Age* (New York: The Museum of Modern Art, 1968), 216.
23. Garmire was mentored at CalTech by none other than Charles H. Townes, laser pioneer. Excited by the possibilities of the creative applications of the laser, Garmire was eventually taken into the E.A.T. fold by Billy Klüver.
24. L*aser Light: A New Visual Art*, 12.
25. Schulman, J. Neil, Appendix: "Violet: House of the Laser," in *The Rainbow Cadenza: A Novel in Logosata Form* (New York: Simon and Schuster, 1983).
26. The seeds of this movement can also be found in Stan VanDerBeek's 1965 *Movie-Drome,* an immersive, multimedia art installation in a recycled grain silo. It was envisioned as a communications system or "experience machine." See Gloria Sutton's *The Experience Machine: Stan Vanderbeek's Movie-Drome and Expanded Cinema* (Cambridge: The MIT Press, 2015).

27. US Patent #5443036 A.
28. US Patent #5443036 A, 3.
29. US Patent #6701872 B1.
30. SETI (Search for Extraterrestrial Intelligence), is a non-profit organization formed in 1984. Its members use earthbound, scientific modes of monitoring (radio, telescope arrays) to capture communication attempts from other worlds. Charles Townes was once associated with SETI.
31. Schwartz, R.N. and C.H. Townes, "Interstellar and Interplanetary Communication by Optical Masers," Institute for Defense Analyses, *Nature: Journal of Science* (London: Macmillan Journals Ltd, April 15, 1961), 205.
32. Schwartz & Townes, 207.
33. "Making the Perfect Searchlight," *New Scientist* (London: New Science Publications, May 12, 1960), 1185.
34. Reddit, the popular Internet community, first drew attention to the image on NASA's Jet Propulsion Laboratory website in 2013 with the comment thread "Mars Rover = $800m, Team to Operate = $1b. Drawing a penis on the surface of another planet = Priceless." The volume of online views led the NASA server holding the image to temporarily crash.
35. *The Day After Roswell* was written by Philip Corso and William Birnes as a populist exposé of this theory (New York: Pocket Books, 1997).
36. Riordan, Michael, "How Bell Labs Missed the Microchip," *IEEE Spectrum* (New York: Institute of Electrical and Electronics Engineers, 1 December 2006). <spectrum.ieee.org/computing/hardware/how-bell-labs-missed-the-microchip>
37. The Henry Ford Museum, accession file #63.173.1.

HELLO MY NAME IS VOTRAX

> 100 PRINT#-2, "Hello my name is Votrax"
> 200 A$= "That is incorrect": PRINT #-2, A$
> —Votrax Instruction Manual

> Through constant decay /
> Uranium creates the radioactive ray
> — Kraftwerk, "Uranium" (spoken through a Votrax)

> "It's the sorest throat on record, bonecake-dry."
> —Dave Tompkins, on the Votrax, *How to Wreck a Nice Beach*

A dark theater, empty save for the seats filled by myself, writer Dave Tompkins, and my partner, Bernie Brooks. It's 2014 and we're in the Anderson Theater in Dearborn, Michigan. Paul Elliman is a grey-shirted shadow onstage, hand held above his eyes to shield against the overhead can lights. He peers into the empty venue and asks, "Is there even anyone out there?" He starts up a recording of a Votrax SC-01 circuit talking itself back to life. A breathy rumble rises from the deepest surface of its chip-gut. Breathy aspirations build into layers. Then come the exasperated moans and digital drones. Next, a cycle of root phonemes and dipthong sequences: an "ehhhh" and an "eh" and an "eeee" and a "zzz" and a "schhhh." It tentatively tests its ability to string the sounds together with a quick singsong of vowels. The SC-01 then laboriously stumbles over the shapes of words as it remembers how to speak. It's the deep demon growl of a slowed-down record catching speed and stumbling across the broken segments of its first word: "aa—a—bom-in—nuh-nation…"

If time and space were to overlap, Elliman would be standing in or near the spectral shoes of Jack Morton, the engineer and manager of Bell Labs' transistor development project. (In 1963, Morton lectured from the same stage while presenting an optical maser.) The transistor is, in fact, a key component of the integrated circuit that powers the lungs of the Votrax.

The Votrax Type 'N Talk is a text-to-speech synthesizer invented in 1970 by Richard Gagnon, a computer engineer at Federal Screw Works in Troy, Michigan. Working in his own basement laboratory, Gagnon eventually delivered a viable prototype back to his employers, who were impressed. He suddenly found himself heading the newly formed Vocal Interface Division. While his efforts earned him a promotion, Gagnon's motivation for developing the Votrax was always personal. The synthesizer could dictate words on a computer monitor, which helped ease the burden caused by his own failing vision.

Using technology, the Votrax re-humanized the human voice in a realm that was always hungry for dehumanization. Engineers like Gagnon wanted to produce smoother-sounding synthetic voices capable of realistic expression, natural slides between syllables, and the ability to convey emotion. To do this, Gagnon exploited the use of the phoneme—the most basic building block of sound, assembled together piece by piece to form human speech—and modeled the output to mimic his own vocal cadences.

Human speech resonates through frequency bands called formants. The Votrax exploits this acoustic energy with a filter that acts like a copycat for the human vocal tract. Hertz by hertz, sound and its frictions excite the Votrax circuits; the phoneme sequences are triggered and "talk back" according to pre-programmed rules. Adjusting for this kind of man-machine interface sometimes required the words to be spelled phonetically:

> circuit became cirkit
> generator became generayter
> machine became mosheen
> radio became radeo

A demo tape of the Votrax Type 'N Talk system, which is stored at the Smithsonian archives, runs through sets of dull phrases, known as Harvard Sentences: "Rice is often served in round bowls. The juice of lemons makes fine punch. A box was thrown beside the parked truck. The hogs were fed chopped corn and garbage." These neutral, nonsense sentences are essentially the equivalent of the SMPTE color bars of television broadcasting or pangrams such as "The quick brown fox jumps over the lazy dog," which is often used to display typefaces or test keyboard equipment. Harvard Sentences are phonetically balanced phrases designed to test the range of the Votrax when transmitted over telephone lines.

In 1980, the company produced the first integrated circuit speech-synthesizing chip—the SC-01—which is present in the Votrax Type 'N Talk model. The LVM-80 model could "'speak' words and phrases with a quality that is said to be 'virtually indistinguishable' from that of the original speaker, whose voice is previously recorded, digitized, and stored in the form of individually addressable messages." Two knobs on the front allowed a user to adjust volume and frequency, and create louder and faster-speaking voices with higher and lower pitch.

Video games owe much to the SC-01 chip. The Midway and Gottlieb companies used it to create sound effects for the pop bumpers and drop targets on the playfield of pinball machines like Black Hole, as well as arcade games like Wizard of Wor, Gorf, and Reaktor. When Q★Bert (known in its earliest phase as "Snots & Boogers") lets loose with a foul-mouthed curse of "@!#?@!," those synthesized swears are Votrax-powered.

The Votrax even found its way onto prime-time television with a commercial for the Parker Brothers board game of Q★Bert. A friendly rapping female voiceover runs alongside live-action clips of full-grown humans stuffed into Q★Bert, Coily, and Slick costumes as they bounce around an Escheresque pyramid of cubes: "The card game's sold separately, that's a fact—so is the board game, that's a rap!" By 1983, the year the oversaturated video game industry suffered a brief swan song, Gagnon could be found commuting to Vocal Division HQ in a car marked with a Michigan "VOTRAX" vanity plate.

The Votrax also talked its way into Atari 1400-series computers, which sold poorly, as well as the TRS-80 and other off-brands by RadioShack and IBM. Programmers who didn't have the stamina to read their programs in monochrome all day used the Votrax to narrate their screens. Companies licensed the SC-01 chip and made their own Votrax knockoffs: Audibraille, the Real Talker, Digitalker, Total Talk, Handy Voice, the Mockingboard, and the Alien Group Voice Box. The Heathkit build-it-yourself electronics company, based in Saint Joseph, Michigan, gave their HERO 1 and Hero Jr edutainment robots a voice by using the SC01 chip. HERO could guard its master's bedroom door, and if its motion sensors were activated, it might utter "SOME-THING-MOVE" in a monotone, in the night. During the day, if no commands were programmed, the robot would default to roaming the house looking for humans, keeping its own company while chatting away in "Roblish."

It wasn't all frivolous robots and teenagers tapping keys in dark spaces with the Votrax. In 1974, at Michigan State University's Artificial Language Laboratory, a Control Data 6500 computer nicknamed "Alexander" was paired with a Votrax synthesizer module. The goal of MSU's unit was to create computer speech aids for people with vocal paralysis and communication issues. One such person was Donald Sherman, who suffered from Moebius syndrome, a rare neurological disorder that causes facial paralysis.

In early December, using Alexander as a sympathetic vocal tract to lean on, Sherman dialed Domino's Pizza. But the calltaker assumed they were being pranked, and hung up twice. The now-defunct Mr. Mike's Pizza was a success on the second try, as Sherman phonetically typed his message on the keyboard, and the Votrax made his demands:

> "I am YOO-sing a special de-VICE to help me to com-MYOO-nicate. ... PLEASE be PA-tient while I pre-PARE my re-SPON-ses."
> "O.K., we didn't understand what was going on here, it's just—" [Alexander cuts him off]
> "I'd LIKE to OR-der a PEE-tza."

Forty-five minutes, seven dollars, two pizzerias, and four calls later, Mr. Mike came through with a pepperoni, mushroom, ham, and sausage pizza. This moment goes down in the books as the first time a computer speech synthesizer was used to assist in a commercial transaction. Hoping to seek forgiveness for their earlier rudeness, Domino's apparently caters events at the MSU Artificial Language Lab today.

In 1985, if you dialed a phone number in two-thirds of the United States and found yourself stalled in directory-assistance hell, chances are your earpiece would deliver the Votraxified voice of Kathleen O'Brien, parked somewhere in telecommunications limbo on the Votrax Incorporated Audio Response System. Her mechanized woman-machine cadence, an early example of "phonecasting," would reassemble single digits into information: telephone numbers, bus schedules, stock reports, lottery numbers. O'Brien fell into the job while working as a copywriter at an agency that also happened to have Votrax as a client. A former model with flawless diction and vocal modulation, she had the perfect voice for computerization. Four other voices—including two male voices—were considered for the ARS device, but O'Brien's was so popular, it might as well have been the only one. John Lauder, the owner of Votrax, mused, "Female voices, for whatever reason, are preferred over male voices."

Case in point: Advertisements for text-to-speech synthesizers in the 1970s & 80s (Votrax or otherwise) amplify dreams of technologized gender, using full, feminine lips and perfect toothy grins, as if to deny the reality of the guttural, mechanical talkback of computer voices at the time. Two back-to-back films released in 1979 and 1980 depict women as omniscient radio goddesses. Lips on full frame, voices pressed into microphones. In Walter Hill's *The Warriors*, an unnamed female DJ (Lynne Thigpen), who is filmed from the nose down, smooth-talks "boppers" through the night and antagonizes the anti-heroes all the way home to Coney Island. In *The Fog*, DJ Stevie Wayne (Adrienne Barbeau) owns her lighthouse radio station outright, playing smooth jazz for night fishermen. When the ghost ship *Elizabeth Dane* arrives with its leprous zombie-sailors in a luminescent green cloud, Wayne is marooned in the tower and hunkers down to guide townspeople to safety over the airwaves.

The Digital Equipment Corporation produced the DECtalk

system in 1984. This text-to-speech synthesizer functioned in a similar way as the Votrax. A user phonetically spells out words and the machine reassembles them into passable speech, based on its bank of internal phonemes. An added bonus was that you could control the enunciation inflections of the machine and force it to sing. Also of note: The DEC was the same company that created the PDP-8, which appeared as the techno-protagonist in Nigel Kneale's 1972 television play *The Stone Tape*. And the bones of a modified DECtalk and a Speech Plus (model CallText 5010) module had each been in use for years by theoretical physicist Stephen Hawking. In the early years, Hawking's voice was filtered through the Gagnon bandpass—Hawking's voice was Votraxified too, for a time. After seeing Hawking give a speech through an amplified Votrax, the author Allucquère Rosanne Stone wondered "Where does he stop? Where are his edges?"

The DECtalk gave users the option of nine built-in voices, including Beautiful Betty, Huge Harry, Kit the Kid, Rough Rita, Perfect Paul, and Whispering Wendy. On closer consideration, this device is perhaps a readymade study in the sonic spaces of gender politics, meshed with techno-optimism. Some of the female voices were formed by simply scaling up the frequencies of the male root voice.

- *I am the standard male voice, Perfect Paul.*
- *I am Beautiful Betty, the standard female voice. Some people think I sound a bit like a man.*
- *I am Huge Harry, a very large person with a deep voice. I can serve as an authority figure.*
- *My name is Kit the Kid and I am about 10 years old. Do I sound like a boy, or like a girl?*
- *I am Whispering Wendy, and have a very breathy voice quality. Can you understand me even though I am whispering?*

———

In the early 1980s, the first round of "speaking" video games were released, including Stratovox ("Help me!" "Lucky!") and Space Spartans ("The battle is over!). But these claims were false advertising, because the games were using the craggy impressions of the human voice via speech synthesizers like the Votrax and DECtalk, which formed phonemes into words, rather than using the human voice itself.

A different kind of demanding female voice emerged with Suzanne Ciani's work on the 1980 pinball machine Xenon. This is one of the earliest moments when the feminine voice was fully sexualized onto a silicon chip for a mass audience. Using her "Voice Box" synthesizer, Ciani played up the trope of the immodest orgasm and became the first female voice in game history.[1] The original PinBot's husky call invited players to "enter Xenon" and "try tube shot." It emitted satisfied "ohs" and "ahhs" as quarters were dropped in. (She also downshifted her voice to become the male voice in the game.)

Ciani milked all she could of the mere seconds of memory available on the ROMS plugged into the game's Sounds Plus Card and the Vocalizer circuit board (later replaced by a "Squawk and Talk" module). One pinball collector in a YouTube review proclaimed Xenon "the greatest ass in pinball" and the "trophy wife of pinball games." "You don't want to play too much," he continued, "you just want to show her off and give her drinks so long as she keeps laughing at your jokes." Apparently, the Gamergate attitude also applies to pinball.

Circling back to the Votrax, this device evokes the original wavelengths of Gagnon's zombie vocal folds perpetually opening and closing, the real man himself now anonymous and forgotten. The full vocalprint and nasal resonance of Gagnon's speech is trapped inside various Votrax technologies—a readymade

---

1 *A Life in Waves*, Directed by Brett Whitcomb, performances by Suzanne Ciani, Window Pictures, 2017.

haunting of formant energy, waiting to holler back at us, over four decades removed from its source.

Decades later, the Votrax-obsessive Paul Elliman tracked down Gagnon's daughter. She wrote about the appearance of her father's voice in the music of Kraftwerk: "I can recognize it immediately. ... I have an ear for his voice. ... The Votrax is my father, if that makes any sense."[2] Gagnon's voice lingers as what Erik Davis would call an "analog doppelgänger, spectral distortion, or a vocal ghost lingering in imaginal space."[3]

The true story goes that Florian Schneider, a founding member of Kraftwerk, discovered the Votrax in Detroit in the mid-1970s while on tour for Autobahn. Schneider, a "collector of artificial voices," and his bandmates were the first to harness the robotic drawl of the vocoder.[4] And the Votrax became one more filter in the archive. Gagnon could be an honorary member of Kraftwerk, as their German VS6.G2 version of the Votrax—sometimes filtered through a vocoder—staggers its way through "Uranium" and chants out multilingual "Numbers" on the 1981 album Computer World. It can be heard on "It's More Fun To Compute," "Musique Non-Stop," and "Techno Pop," as well.

The fragments of Gagnon's voice appeared not just on Düsseldorf electro tracks, but spilled into techno and early hip-hop, as well. They were cemented into the sonic landscape of Detroit radio by Electrifying Mojo, and captured in a grainy

---

2 Elliman, Paul. "Detroit as Refrain." *The Serving Library, Bulletins of The Serving Library*, no. 8 (2014). <www.servinglibrary.org/journal/8/detroit-as-refrain> Accessed 5 March 2016.

3 Davis, Erik. "Roots and Wires Remix." *Sound Unbound: Sampling Digital Music and Culture*, edited by Paul D. Miller, (Cambridge, MA: MIT Press, 2010), 63.

4 Tompkins, Dave. *How to Wreck a Nice Beach: The Vocoder from World War II to Hip-Hop, the Machine Speaks* (Chicago, IL/Brooklyn, NY: Stop Smiling Books/Melville House Publishing, 2010), 185.

but beloved YouTube clip salvaged from the Detroit cable TV program The New Dance Show. In the clip, recorded at Club Studio III in 1991, a woman opens the dancefloor with an impressive launch off a stage, landing in full splits (in heels, no less); the Detroiters form a regimented dance line, then bust out and take turns vamping down the aisle to Kraftwerk's "Numbers." Kraftwerk's impact on Detroit native Rik Davis, one half of the proto-techno duo Cybotron, is undeniable. However, when asked if he was aware of the device while recording the classic 1983 record Enter, Davis responded, "Votrax is a no-no."

If you powered your computer down before turning off the Votrax, it would scream a death knell of "AAAA—RRRR-GGGhhhh..." But this faux sonified death gasp is all smoke and mirrors because the Votrax always comes back to life. It's a loosely defined testament to the Votrax's breakout desire to sing. As the advertisement for the Votrax goes, "Now You're Talkin'!"

FETCH X
(Living Ghost Lifts Up The Receiver)

> She darkened my path, like a troubled dream,
>   In that solitude far and drear;
> I spoke to my child! But she did not seem
>   To hearken with human ear.
>
> She only looked with a dead, dead eye,
>   And a wan, wan cheek of sorrow;
> I knew her fetch! – she was called to die,
>   And she died upon the morrow.
>
> —John Banim, 'The Fetch', 1825

I. *X Reports*

She didn't mean to cause a haunting.[1] She was, however, well aware of how the haunting came to be. Some words that define that particular time: "malignant," "lost," or, in more paranoid times, "premeditated." A phrase might be better: *The wooden anchor would hold if only it were larger, thinks the fool.* Part of her would always remain there, in that cobblestone cottage on the

---

1 E422.1.1.1. *Two-headed ghost.* Swiss: Jegerlehner Oberwallis 334b s.v. "Geister"; England, Scotland, U.S.: Baughman.

shores of the shallow lake, where shipwrecks lined the bottom, buried in silt.[2] It was important she leave that place. Cutting ties was traumatic, and so a part of her remained behind. For a time, she became a living ghost, anchored to that back room by the shock of white that cropped up in her hair soon after. She had not left soon enough to prevent this.

She became suspicious when reports from her ex began to trickle in. He asked leading questions skirted around larger questions that he didn't have the nerve to ask directly: "Have you been having dreams about the cottage lately? Have you had any of the old death dreams? When you dream, do you wander far?" But never, "I'm sorry."[3]

Before he called to ask these questions, it was obvious he'd attempted exorcisms of his own. He burned sweet sage and tobacco and slept surrounded by white candles.[4] She was made to feel her claims of innocence were selfish. Apparently, she had not made clear the urgency of her state of illness before slipping into a coma on the bed for three days.

Haunting a house was not something she'd ever intended to do, especially since she was undeniably alive. Years later, when she asked about those missing days, he insisted she'd been up out of bed, walking around the house, albeit as a ghost.[5]

Things from that time began haunting her, too. In that purple limbo state, she'd unconsciously wrapped her body in a

---

2 E711.7. *Soul in stone.* (Cf. †E761.5.5.) Irish myth: *Cross; Icelandic: Boberg.

3 E461. *Fight of revenant with living person.* (Cf. †E261.1.3.) [...] N. A. Indian (Teton): Dorsey AA o.s. II (1889) 150, (Passamaquoddy): Leland Algonquin Legends 349, (Seneca): Curtin-Hewitt RBAE XXXII 96 No. 7.

4 E694.2. *Frustrated woman reborn as tobacco plant.* India: Thompson-Balys.

5 E175. *Death thought sleep. Resuscitated person thinks he has been sleeping. He exclaims, "How long I have been asleep!"* *Köhler-Bolte 555; Wesselski Märchen 192.—India: *Thompson-Balys; [...] S. A. Indian (Yuracare): Métraux BBAE CXLIII (3) 502.

white sheet cross-stitched with deer and flowers by her great-grandmother. It suggested a burial shroud. At some point, probably on the second day, she expelled the yellow bile of an empty stomach onto it.[6] Although it was her favorite linen, she could never touch it again. The sheet became tabooed. Before realizing this, she made the mistake of putting it on her bed one night, and when she touched it to peel back the covers, she felt the electric shock of the near-death experience return.

II. *X Loop*

She didn't realize the gravity of the situation until her mother called in a panic. She was returning a call she received from an electronic Otherplace. She lay in bed upstairs, unconscious, experiencing vague dreams about making a phone call to her family. A piece of her split away and found its way into the telephone wires.[7] She slipped through the phone lines toward her family. She thought about making all of the telephones in their house ring louder, and longer, but the sound on her mother's end announced itself like bells buried under cotton. When the line clicked open, her repeating loop began:

> "Hello…hello…hey…It's very important that you call me…"

It was her crisis apparition calling. This narrative repeats itself as a trope: *phone calls from the dead.*

---

6 E261.2. *Dead arises when shroud bursts and pursues attendant.* Estonian: Aarne FFC XXV 113 No. 3; Lithuanian: Balys Index No. *369; Finnish: Aarne FFC XXXIII 39 No. 3; Lappish: Qvigstad FFC LX 40 No. 4.

7 E721.1.2.4. *Soul of sleeper prevented from returning to his body when soul as bee leaves body and enters hole in wall beside which he is sleeping.* (Cf. †E734.2.) England: Baughman.

"I am telephoning you—telephoning to tell you and to warn you of X."

The contemporary legend of the "telephone call from the dead" will often involve an informant reporting, for example, a favorite uncle lying comatose in a hospital bed, reaching out in the moments before his death by way of a psychic collect call. Through the strained muddle of the line, the informant claims to recognize the uncle's distinct timbre from a time when he was healthy, yet with an added layer in his voice indicating the panic (or resolve) toward his impending death.[8] Simple but charged words, relayed:

"Hello, dear. Don't forget to put the X away."

Of course, temporary amnesia blocked the receiver of the call from realizing that the conversation was an impossibility until the line went dead or the voice faded away. Fainting of informant sometimes occurs here.

---

8 E417. *Dead person speaks from grave.* Madagascar: Sibree FLJ I 202ff., Larrouy RTP IV 305.

III. *Fear of X*

Mrs. B., wife of a doctor in good practice and general esteem, looking towards the window from her pillow, was startled by the appearance of her husband standing near the table …. Now, the living and breathing man was by her side apparently asleep … After gazing on the apparition for a few seconds, she bent her eyes upon her husband to ascertain if his looks were turned in the direction of the window, but his eyes were closed. She turned round again, although now dreading the sight of what she believed to be her husband's fetch, but it was no longer there.

Next morning, Mr. B., seeing signs of disquiet on his wife's countenance while at breakfast, made some affectionate inquiries, but she concealed her trouble … Meeting Dr. C., in the street … he asked his opinion on the subject of fetches. "I think," was the answer, "and so I am sure do you, that they are mere illusions produced by a disturbed stomach acting upon the excited brain of a highly imaginative or superstitious person." "Then," said Mr. B., "I am highly imaginative or superstitious, for I distinctly saw my own outward man last night standing at the table in the bedroom … I am afraid my wife saw it too, but I have been afraid to speak to her on the subject."

— Patrick Kennedy, 'The Doctor's Fetch', 1891

IV. *X Unbound*

The fetch is a living ghost, a roaming bi- and multi-local double of the self. It appears in folklore of the British Isles as a harbinger of death, and sometimes (though very rarely) as an indicator of a long life. To see one's fetch is to see an omen of mortality, to witness one's own "giving up the ghost."⁹ In Scotland, the fetch is known as a wraith, the *taradh*—or as *manadh nan daoine bko* (the omens of living man). It is called the shade or double in Britain, and the double walker or *doppelgänger* in Germany. The fetch is a signature, an authority of presence in a place where your living body is not.

Their form shifts and is made of shadow. It reconstitutes itself in the memory of the beholder in order to be recognized. An eerie glow emanates from within. A black glow. The fetch reflects and refracts light, an aesthetic akin to the idea that all light that has *ever* existed is drawn toward it and deadened by it. Abnormal light defines the fetch as *something else*, something recognizably *not present*.[10]

Her ex's inquiries of wandering dream states concerned her, because she had always had her suspicions that she was walking and muttering half-lucidly someplace else. She often woke up exhausted, her throat scratched hoarse. So she took proactive measures to reclaim her personhood, to reunite her distributed body with her fetch. The anthropologist Alfred Gell speaks of the notion of *exuvia*—unbound pieces of oneself, purposefully cast over a large geographic space as an act of power. By distributing the physical outgrowths of the body—hair, nails, teeth—one gains the power to psychically travel, and to control. This was not her intention. The more she and her fetch became confused, the more confusedly they entwined, and the more intense the activity during her waking hours became.

The living dead girl once interrupted Christmas dinner when she announced her name before sitting prim and semi-

---

9 E530.1.5. *Ghost light indicates impending calamity.* Scotland: *Baughman.

10 E422.2.4. *Revenant black.* Irish: Cross, O'Suilleabhain 63, Beal XXI 324.

transparent by the stone fireplace.[11] Sometimes the bed would indent under her feather weight as she sat upon the bed, just perceptibly enough for him to question the dependability of his senses. There was no question that he could hear wheezing out in the hallway. The sound of clothing, like the swishing of a wool skirt, came from the direction of the guest room and toward his door, over and over again. She would pass over his inert body, retreat, and return again.

Other times, he would wake up in the middle of the night and it would become clear that the lull of the surf breaking was not the lake, but the sound of rhythmic breathing, hovering five inches or so above his face. Wafts of her scent filled the room. He could feel the air move with each discarnate breath, but he could not move himself. He attempted the sign of the cross with his tongue, but that ward failed him because he was too afraid to be faithful. Body parts distinct from a body—a forearm or a hand, a thigh or a single lock of hair, would present themselves as accusations in dreams.[12] The knocking and creaking of the room, much like a bow of a ship, would sound itself as a distress call. He learned Morse code and deciphered the signals.[13] They mirrored her crisis phone call:

> *"...terribly important...call me...hello..."*

How else could it be?

She burned sweetgrass and salt, too, of course, but it wasn't strong enough. So she washed her floors with Buffalo Ammonia.

---

11 E386.5. *Light remark about what person would do if ghost appeared causes ghost to appear.* (Cf. †C10, †C13.) England: Baughman.

12 E422.1.11.1. *Revenant as an eye.* U.S.: Baughman; E422.1.11.3. †E422.1.11.3. *Ghost as hand or hands.* England, U.S.: *Baughman.

13 E533.1. *Ghostly bell sounds from under water.* England: *Baughman.

She lit candles and had others light them for her, too: Cast Off Evil, Fear Not to Walk Over Evil, Fiery Wall of Protection. She bathed in charmed bath crystals: Banishment, Go Away Boy, Come to Me. She burned incense: Clear Away That Evil Mess, Run Devil Run, Psychic Uncrossing. She went so far as to work a Break Up spell onto herself.[14]

The cottage was built in the late 1880s as a summer getaway, guided by the hand of his domineering grandmother. Later generations weatherproofed and lived there year-round. In the dead of winter, the lake would freeze solid. At the end of the season, when everything was perfectly aligned, the tension would come to a head on the lake and the ice would explode into shards of tall floes.[15] The sound was booming, followed by crystalline tinkling, and it would wake the whole house. The night she became a fetch was not accompanied by such natural phenomena.

He considered bricking up the windows of the bedroom and paneling over the door. It would take a lot of work, since the room itself was bordered on three sides by paneled glass casement windows. Best to brick the whole thing in—that room was always cold, anyhow.[16] It would remain concealed until a family member found the room from the exterior of the house years later, looking up to the jutting addition and realizing no pathways led there.[17]

A bricked-up window is a window to be looked upon with suspicion.

---

14 E412.3.2.1. *Ghost asks to wash his shirt.* Lithuanian: Balys Ghosts.

15 E437.1. *Revenants banished to glaciers and uninhabited places.* Swiss: Jegerlehner Oberwallis 296 No. 27.

16 E501.17.5.3. *Wild hunt avoided by keeping in house with windows closed.* (Cf. †E501.14.6.) Plischke 76.

17 N517.1. *Treasure hidden in secret room in house.* England, U.S.: *Baughman.

He weighed his options. He didn't want to defile his grandmother's pride in the home, but he knew it was inevitable that it would be haunted by the "phantasm of the living" who continued to loiter there. There were many other rooms in the cottage, and so he carried out his plan. He left the fetch to find its way home.

Over the course of two days, she began to feel an increasing sense of suffocation, and couldn't understand why. This carried on for several years.[18]

One day, while searching her own name on the Internet, she found her own obituary in the form of a memorial photograph. A friend found it too, and when he couldn't find traces of her life online, he assumed she'd been dead for many years.[19] It was an image of her favorite cat, with the byline:

"In memory of X."

The image and the X., bound to her, were to blame for this misunderstanding. There were people walking around, believing she had died, and mourning for her.[20] Finding her fetch in the digital ether was not something she'd accounted for. All of those traditional means of banishment and reunion were for naught. Her body had been floating through the phone lines, and he grounded it via the dial-up modem in his room. She was the source of the haunting, but he had caused its continuation.[21]

She called him with the solution, annoyed at his stupidity and private apologies directed everywhere but toward her. She

---

18 E599.1. *Ghost searches for breath.* U.S.: Baughman.

19 E532. *Ghost-like picture.* U.S.: *Baughman.

20 E402.1.6. *Crash as of breaking glass, though no glass is found broken.* England, U.S.: *Baughman.

21 E533.2. *Self-tolling bell.* England: Baughman.

asked him to remove the image. He refused, but he did remove the byline. It was enough to break the connection.[22]

V. *Fetch X.*

She lit my path, in a carefree dream,
  In that empyrean far and revered;
I spoke to my child! But she did not seem
  To hearken with human ear.

She only looked with a quizzing eye,
  And a fateful cheek of sorrow;
I knew her fetch! – she was called to die,
  But would live the long, long morrow.

—John Banim, 'The Fetch', 1825, False Variant

---

22 E402.1.7. *Ghost slams door.* Canada: Baughman.

IV. BROADCAST

RUSH IN REVERSE

One of K.'s first memories was of the night she screamed herself awake, pounding the scratchy brown and orange plaid couch in the living room with her fists and feet.[1] A rented laserdisc machine was playing Sam Raimi's *Evil Dead*. The cabin-rush scene had influenced her edges of sleep into becoming a nightmare. This violent resonance spooked her two brothers, who, up until that moment, were sitting cross-legged (and likely stoned) in front of the floor model RCA. They promptly scrambled to their feet.

A year or so later, at the same rented house, the entirety of its downstairs was crisped black by an electrical fire.[2] After the fire, a trip was made to the house to recover important documents—marriage and birth certificates, and the like. K. was told by her mother and aunt to wait on the porch, but she wandered into the living room anyway. Her nose filled with the smell of cinders and her field of vision was overtaken by the uniformity of half-familiar shapes turned carbon black. Just so black.[3]

---

1 E34 *Resuscitation with misplaced head.*

2 A689.1. *Dark puddles in hell.* Irish myth: Cross.

3 E593.3. *If no lamp is lighted in a house for a period of fourteen days, ghosts take it for their dwelling.* India: Thompson-Balys.

In a family photograph, now lost, is an image of K. sitting in a high chair at the dining room table, a backdrop of classic 1970s yellow, orange, and green floral wallpaper. A diagonal blast of sunlight across the wall behind her. Unexplained tendrils of ghost smoke free-floating in between the wall and her chair. Perhaps a forecast of the future.

Posters of 1980s horror films hung on the slanted cove of her brother's bedroom walls. The *Evil Dead* demon hand was among them, shooting up to steal the world from beneath the earth. The fire didn't reach the upstairs bedrooms, but the scent of smoke damage and the Freon that leaked from the super-heated refrigerator saturated everything they took into their new house.[4]

While visiting London as an adult, K. once felt the embedded panic of that cabin-rush scene from *Evil Dead* return with no real cause as she opened a hotel-room door.[5] The sense of an ominous gale blown straight up the steep, winding stairs and through the glass-and-wire-mesh safety door. The burn of a less-than-friendly slap planted square on the back knocked her into the room.[6] Backing out of the room, she saw nothing but an empty hallway. Nothing except for Zeze, the hotel cat, watching from a windowsill down the hall, judging K.'s reaction and yawning as if the manhandling of hotel guests by invisible hands was all just *so* old hat.[7]

---

4 Z13.4 (j) >> *Man chased by coffin, which follows him.*

5 E272. *Road-ghosts.* (Cf. E332ff., E582.) Ghosts which haunt roads. Lappish: Qvigstad FFC LX 40 No. 10; England, U.S.: *Baughman; New York: Jones JAFL LVII 248; North Carolina: Brown Collection I 674.

6 M118. *Swearing on a skull.* Irish myth: Cross.

7 B181.1.1. *Cat with remarkable powers of sight.* India: Thompson-Balys.

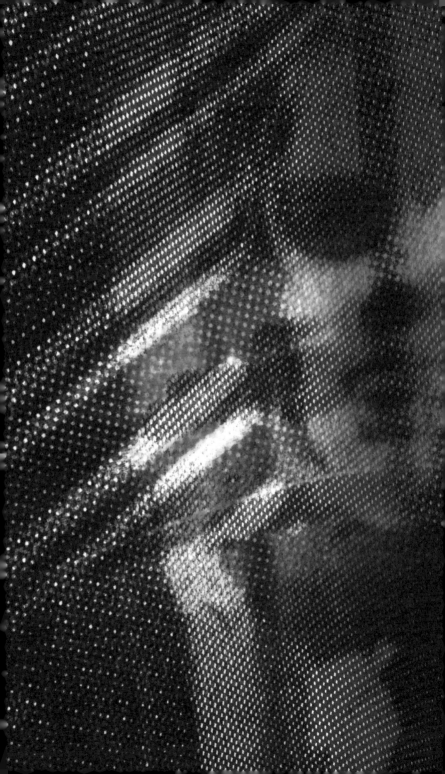

WE HAUNT YOU WHILE YOU WATCH
TV Hijackings, The Eerie In The Middle, And Cochlear Control

I. *Ears That Ring Like Test Tones*

I couldn't tell you exactly when it became clear that my ears became scrambled. I assumed, at first, that the high-pitched whine I heard, intermittently and unpredictably, was coming from somewhere in the house. Perhaps a lost signal from the old wires in the black mass of the basement. Or clairaudience brought on by an electrical socket gone rogue.

But driving on the freeway one summer night in Detroit, the squawk rose up in my ear under decidedly non-domestic circumstances, more deafening than ever. The car began to feel as though it was spinning across the crumbling pavement. This loss of ground was wholly internal. I stayed seated and the car stayed within its lane under the steady hands of my partner while I floated out of sync with my body and sideways through my seat, until the sine tone subsided and the world merged back into proper register.

Something else became permanently decoupled that night—the looming threat that a crippling banshee blast from somewhere within my left ear would return. The diagnosis I received of Ménière's disease—commonly the creative person's bailiwick—is an alarming, yet fitting condition for a sonic researcher obsessed with the history of anomalous experiences.

The drop-out of my cochlear microphonics quickly became zombie-like. Several times a day, as the unstoppable hum rose to a high screech and met the threshold of an unknown internal noise floor, my ears became deader, yet more sensitive. Internal worlds amplified, while the external one was filtered out through a whomping deep sub-bass filter that sounded like surfacing from the bottom of the sea.

Describing to others how my day-to-day hearing had changed, I would compare my head to a sound mixing board, with all of the channels spiked into the red. In a diner, the sound of eggs sizzling on the grill were perceptually at the same volume as a conversation across the table. These newfound bat ears were made useless by information overload, and the intense pressure that accompanied the tinnitus felt a bit like having a plunger stuck to the side of my head. The tinnitus was subjective—my body decided to deploy high frequency warfare upon itself. The threat of bio-acoustical failure came with each bout of tape-hiss drop-out and squeal. The space between my ears began to feel like the jumbled guts of a cassette tape, defiled and stretched out to rot in the sun. I asked a friend to put his ear to mine, to see if he could share in the metallic tone my ear was emanating. Apparently, in some extreme cases of tinnitus, this is a possibility, but not in mine.

Ears that ring like a test tone have been an ongoing disruption, well before the ghost of Prosper Ménière drop-kicked his way into my head. That same test-tone sound permeated the latest of late nights of a 1980s childhood, a key bit of aural memory in that futile game of testing the boundaries of what staying up late *could mean*. It usually meant falling asleep on the couch.

Once, I was upturned from slumber by the panic-inducing ring of Klaxon horns as the nuclear bomb dropped in Raymond

Briggs's *When the Wind Blows*.[1] This particularly scarring revelation sparked a lifelong cycle of apocalyptic dreams and the fissionable connection to anxiety that the emergency snow horns in my neighborhood produce, even today. There was also the utter dread of waking up to the Canadian national anthem playing over the television. Somehow being met with bombastic pageantry always seemed more menacing than whatever late-night horror film or underground music program I'd been watching.

There was a very specific kind of sleep disturbance linked to the era when television stations habitually signed off at night, when the last hurrah of the channel invaded the beginnings of dreams. Whether it was an anthem, the recitation of a poem, city skylines, a montage of natural scenery, scenes of broadcast workers and their equipment, children's choirs, or animations of sleepy cartoon men locking up the house before getting ready for bed—every geographic area had its own unique television sign-off. Consequently, every locale had its own particular intrusion into the thick dream cycles of accidental armchair sleepers. This was an absolute bookend to the day—a signpost that read, "You have reached the time that that is *beyond late*." The end of the sign-off segment conjured the inevitable arrival of the SMPTE gradient of color bars, weighted with black at the bottom, sounding off with a scream.

II. *The Grinning Horror*

Rebuilding the cradle of culture in November of 1987 finds us deep in the era of Reagan (or, in my case, Brian Mulroney). "I Think We're Alone Now" and "Heaven Is A Place On Earth" ruled pop radio nationwide, while *Night Flight* and *Dr. Demento* were late-night staples of Chicago talk radio.

On November 22, 1987, two Chicago television stations found themselves playing unwilling host to a signal pirate in a Max Headroom mask. At 9:14pm, WGN-TV's signal clipped to black and then flicked back on to reveal a charlatan in a Headroom mask, dancing convulsively in his chair, accompanied by a squelching, distorted soundtrack.[2] Quick-thinking WGN engineers pulled out all emergency stops to regain control of the network within 30 seconds. When sportscaster Dan Roan returned to his report on the Chicago Bears football game, he appeared dazed, and chuckled awkwardly: "Well, if you're wondering what that was... so am I."[3]

Two hours later, at 11:15pm on WTTW-11, during a rerun of the *Dr. Who* episode "Horror of Fang Rock," viewers became outraged by a longer, ruder interlude, which lasted about 90 seconds. The wholesome act of watching public television had rarely felt like more of a violation.

III. *"Be Still Now and Listen"*

"Horror of Fang Rock" originally aired on September 3, 1977, just two months before another infamous anomaly known as "The Southern Television Broadcast Interruption." On November 26, 1977, at approximately 5pm, the UHF television transmitter at Hannington in South East England was overruled for what probably felt like a lifetime (in reality, it was six minutes). A second rogue transmitter, acting as a repeater had latched into the main station and bounced unauthorized sound far and wide over the countryside. It was, in fact, the first known example of an illegal television signal intrusion made by an individual. A watery voice that was run through a flanger made the evening news wave, drip, and roll as it spoke over the anchorman, introducing itself as "Vrillon, a representative of the Ashtar Galactic Command."[4]

In the days leading up to this event, the first nodes of ARPAnet, the network that became the basis for the Internet, were established. France ravaged the French Polynesian atoll of Muruora by detonating a nuclear explosion. Ignited in an underground shaft drilled deep into the atoll's volcanic rock, this was neither the first, nor the last, of France's nuclear tests in the South Pacific.

While ARPAnet's future impact was yet to be understood, the public was primarily concerned with protesting the proliferation of nuclear arms. As the voice of the Southern Television event, "Vrillon" played the part of a stern advisor within this uncertain landscape, spreading a message of peace and acting as a harbinger of global destruction. "Be still now and listen," he said, "for your chance may not come again. All of your weapons of evil must be removed. The time for conflict is now past and the race of which you are a part may proceed to the higher stages of its evolution if you show yourselves worthy to do this."

Looking back at the events of 1977, humanity seemed hungry to jump ship to better planets—or at the very least, actively reveled in the fantasy of alien invasion. Between the launch of the Soviet's Soyuz 24 and NASA's Voyagers 1 and 2, there was the release of *Star Wars* and *Close Encounters of the Third Kind*; the SETI project's reception of the deep-space "Wow!" signal;[5] and reports of mysterious, luminous objects in the sky over Russia, Finland, and Denmark, which became known as the Petrozavodsk phenomenon. People were suffering a new kind of celestial whiplash from all that neck-craning toward the stars.

With the departure of Vrillon's voice and the righting of ITV's signal, the 1943 *Merrie Melodies* cartoon "Falling Hare" emerged, featuring Bugs Bunny lazing around a US Army airfield, chawing on a carrot and directing viewers to a page in a book about those diabolical creatures called "gremlins" that sabotage airplanes. As Bugs erupts in his best maniac laugh, a tiny gremlin with a giant sledgehammer begins to whack the bomb he's sitting on. Twenty years after Bugs ate gremlin humble pie, on a 1963 episode of *The Twilight Zone,* William Shatner would lose his mind at 20,000 feet trying to convince a stewardess that an identical creature was destroying the wing of the airplane. Vrillon, in his own way, was the gremlin that first peeled back the layers of transmission technology to expose the precise weak point for jamming up the works. Vrillon's identity, much like the culprit behind the Max Headroom Incident, was never discovered.

IV. *The Eerie in the Middle (Transcription)*

In its interruption of the everyday "normal" of the *Dr. Who* episode "Horror of Fang Rock," the Max Headroom signal-jacking created an accidental media cut-up. In the transcription below, the bookends of dialogue are made more potent by the make-do mischief in between, more portentous than the effect of a disaffected channel-surfer, remote in hand. It is impossible to prove, but pure speculation would assume that few viewers could stop watching the degeneracy of the Headroom clip unfolding, once it began.

[Dr. Who " Horror of Fang Rock"]

LEELA: That is stupid. You should talk often with the old ones of the tribe. That is the only way to learn.

VINCE: I'll get you a hot drink, miss.

LEELA: I could do with some dry clothes more than—

[Cut to Max Headroom Incident]

That does it… He's a frickin' nerd! [*Laughs*]

Yeah, I think I'm better than Chuck Swirsky… Frickin' liberal. Oh, Jesus! [*Unknown*]

Yeah… "Catch the Wave!" [*Screams and moans holding Pepsi can, throws it away*]

"Your Love is Fading!" [*Laughs, takes vibrator or dildo off finger and throws it away*]

Doot-doot-doot-doot… [*Singing* Clutch Cargo *theme*]

I stole CBS! Doot-doot-doot-doot… [*Returns to* Clutch Cargo *theme*]

Ohh… My piles! [Moans while shaking head, as though defecating]

Ohh… I just made a giant masterpiece for all of the greatest world newspaper nerds! [*Moans. This is a reference to WGN's call letters*]

My brother is wearing the other one… It's dirty. [*Puts on work glove*]

That's what you get for "recycled…" [*Takes off and throws glove*]

[Cuts to different view. "Max" is bent over, his naked rear end exposed]

They're coming to get me! Ohhh! *[Moaning]*

Come get me, bitch! [*Yells while woman dressed as Annie Oakley swats his rear with flyswatter*]

Oh, doooo it!

[Cuts back to Dr. Who]

DOCTOR: As far as I can tell, a massive electric shock. He died instantly.

VINCE: The generator? But he was always so careful.

LEELA: It was very dark.

V. *New Wave Pirate*

On the evening news throughout Chicagoland, appeals were made by newscasters, FCC agents, and federal officials for information about the Headroom imposter. A segment on WFLD Channel 32 appeared with a graphic of a television displaying SMPTE color bars overlaid with the text "Air Pirate." Anchor Kris Long described the most disturbing aspects of the second interruption as the "display of a marital aid and a portion of his (or her) anatomy," adding that the offense was sophisticated enough that it "points towards someone with a broadcast background."[6] The segment ends with a stern imprecation that this type of "freelance exercise in public access" would not go unpunished.

On WMAQ Channel 5, Carol Marin gave her report next to a screen graphic of the classic Jolly Roger skull and crossbones.

Marin pulled no punches, declaring that "the video program ended with the video pirate's bare bottom being spanked with a flyswatter, but his punishment will be far worse if he is caught."[7] Apparently, this act would have required powerful equipment to override the station's signal and would have left behind a traceable "electronic signature." Despite the fact that the FCC and FBI immediately dispatched agents to hunt down the source of the signal override, all attempts to track the pirate remain unsuccessful to date.

This copycat emissary of New Wave culture applied several layers of stealth to his shenanigans. His ass might have been exposed and beaten, but his face was safely obscured. The use of the Max Headroom mask by the pirate was far from random—it was an essential part of the make-do set design that framed the intrusion. The corrugated metal sheeting that rocks back and forth behind "Max" obscured his location, but also mimicked the faux-computer graphic animations of the *Max Headroom* television show.

In one portion, when the pirate holds up a can of Pepsi and says "Catch the Wave!" he is referring to the contentious 1980s marketing battle of New Coke vs. Pepsi. (The official, licensed Headroom character appeared in television commercials for this campaign.) When the pirate throws a glove off-screen and says, "That's what you get for 'recycled,'" this is meant as a pun on the relatively new embrace of curbside recycling in the United States. A pair of opaque lens sunglasses—trademark of both the government secret agent and a go-to device for masking criminal identity—are part of the molded latex pirate's mask. Sunglasses may have been an iconic part of the Max Headroom "look" but, in reality, they temporarily alleviated a health issue of the actor Matt Frewer, who suffered from cornea damage due to the intense haptic contact lenses that were part of his character's makeup scheme.[8]

Oddly, reports often referred to the Headroom Incident as an act of "piracy" and never "hacking." The word "hacker" was just entering mainstream culture, but was mostly reserved as a trope to describe elaborate college pranks at MIT—and its computing subculture—and was not yet fully understood by the general public.[9] Today, the term "hacker" has been somewhat perverted into meaning something of purely malicious and criminal intent. It's also been diluted to convey a time-saving "life hack," such as improving concentration or peeling garlic efficiently.

The Headroom Incident, in a fluid, language-based sense, stands as one of the most elusive (and thereby successful) hacking cases of the 1980s. The contemporary hacktivist group Anonymous cloak themselves under the guise of a now-iconic Guy Fawkes mask. In a 2008 anti-Scientology video campaign, an Anonymous actor wearing a Fawkes mask appears against the classic pulsating yellow-and-red-on-black line animation of the Headroom series.[10] The voice used in the Anonymous video was generated by a computerized text-to-speech program, but also adopts the affectations of dropped frame jitters and stutters that were a hallmark effect of the original Headroom television show.

The exaggerated fracturing of Max Headroom's speech was the subject of a newspaper article in 1987, at the height of the show's popularity. Speech therapists were worried that the popularity of the character "may set a bad example for youngsters by encouraging them to stutter. Children are susceptible to authoritative figures," Theodore E. Emery told the annual conversion of the National Stuttering Project.[11] These homages to Headroom, from the Chicago signal-hijacker to modern hacker collectives, allude to a decades-long distrust of the corporate media that continue to dominate our viewing culture. The "real Max" depicted on the television show was actually a hoax—an actor in prosthetics and a stiff suit against a blue-screen—yet was publicized as a computer-generated character. Regardless of this white lie, the actions of the legitimate digital bandits known as Anonymous were inspired to repurpose the aesthetic cues in order to visually bond their goals with the anti-establishment spirit that infused this short-lived program. Headroom's exit diatribe, inspired by a Winston Churchill wartime speech, seems to have infiltrated the attitude of activists today: "We will nev-nev-nev-never surrender."

VI. *Stacking the Signals*

The first Headroom interruption on WGN was a freaky vision made even more horrifying by its distorted audio blast. But viewers lucky enough to catch the second event on WTTW, Chicago's PBS affiliate, might have wished that segment was silent, too. The voice of "Max" was heavily modulated and ran alongside a hollow sine drone. His words and off-key singing dripped into watery, spectral flange and needling moans, skirting the boundaries of accessible language. The fractured nature of his proselytizing was more agitated and disturbing than the stutter of the "real Headroom." The pirate's use of moans, screams, and off-key singing to fill any possible gap of silence caused the audio-processing effects to peak and crumble. It was likely only through seeing replays on the evening news that the viewers who had witnessed the first round of signal jamming could begin to process of what they had seen and heard. This case transcended the concept of the innocent prank—it made hacking "creepy." The seeds of alarm began to germinate, forecasting a future where the predictable banality of television could be overridden by a malevolent broadcast possession.

The Max Headroom Incident was possible from a technical standpoint because its perpetrator could fluently speak the language of signals and broadcast architecture. Officials believe that, given the expertise and expense of the equipment used, the chances of the hijacking being an "inside job" by a disgruntled employee are quite high. Many details remain sketchy. What is known is that the pirate was physically present in the Windy City, and made use of a mobile rig either from the roof of nearby buildings or from the ground. The heart of the mobile setup included a device capable of radiating powerful microwaves that could overrule the normal flow of a commercial broadcast. A rig like this, according to speculators, could cost upwards of $25,000—a hefty expense to shell out in order to be beaten naked with a flyswatter on primetime television.

At WGN, on-site engineers recognized that something was amiss, and they were able to quickly switch over to an alternate transmitter. WTTW's unmanned transmitter was doing its job, over 1,400 feet in the air at the top of the John Hancock Center and Sears Tower. The pirate's first act might have slipped into irrecoverable obscurity, but the length of time (and the disturbing message) that the second hijacking exploited is one reason why the event is occasionally dredged up by pop-culture disciples.[12]

The cunning genius of the culprits is evident when the language of colliding media formats is examined closely. Minute shifts in time indicate that the masked imposter was not filmed live, but had pre-recorded the segment. Distorted "scan lines" are from the original broadcast, indicating that the Headroom Incident first existed as encoded information on a VHS tape. With the tape now likely lost (unless the guilty party were to ever re-emerge), the event itself only exists in secondary and tertiary forms, like magnetized polyester or digital code. Even if the original exists, the curse of the image-degrading condition known as "sticky shedding" is likely to have eaten away at what is left of the image on the tape. Rick Klein of the Museum of Classic Chicago Television has worked to dredge up the best-possible transfer of the Headroom event (and subsequent news reports) from private broadcast archives, resuscitating and sharing it with the public, so they can partake in whatever is left of the magnetized heartbeat of this local event.[13]

Lesser-quality examples of the Headroom hijacking are also viewable on YouTube. The fact that this footage is confined to Internet channels—and not preserved in a museum of broadcast history—is a mixed blessing. To echo media studies professor Lucas Hilderbrand, "YouTube introduces a new model of media access and amateur historiography that, whilst the images are imperfect and the links are impermanent, nonetheless realizes much of the Internet's potential to circulate rare, ephemeral, and elusive texts."[14]

VII. *Phreaking Evasion*

The Max Headroom Incident pulled out all the stops, both visually and aurally. The Southern Television Broadcast Interruption of 1977 was a voice-only hijacking, while two infamous visual interventions from the mid-1980s were solely text-based. In April 1986, "Captain Midnight" took over HBO's satellite uplink to protest the rise of the channel's subscription rates. A color bar test pattern was overlaid with relatively bland text produced by a character generator: "Good Evening HBO from Captain Midnight. $12.95 a month? No Way! (Showtime / The Movie Channel beware.)"

The Captain's true identity was quickly revealed to be John R. MacDougall, a satellite television dealer who believed HBO's actions were contributing to the decline of his business. As a direct outcome of the FBI's investigation into MacDougall, a new law made satellite interference a felony.[15]

On Labor Day weekend 1987, just a few months before the Headroom Incident, the signal for the Playboy Channel was subverted by Thomas Haynie, an employee of Pat Robertson's rival Christian Broadcasting Network.[16] This text-based jam, now known as "The Religious Takeover," implored pornography viewers to repent for their sins and find Jesus.[17] Ultimately, this stunt revealed damning digital traces and led to the capture of Haynie and a trial, much like that of MacDougall. It's ironic that the text intrusion—an act as anonymous as scrawling lewd graffiti in a bathroom stall—would in fact exploit the characteristics of digital typography as a sort of forensic pathway back to its source.

Archived conversation threads from Reddit, the online bulletin-board community, indicate a potential lead and eventual dissolution of a theory regarding the identity of the Headroom marauders.[18] A user identified as "bpoag" claimed to be an avid "phone phreaker"[19] in 1987, and was acquainted with two older brothers from the Chicago suburb of La Grange. "Bpoag" believed the brothers' abilities with electronics and personal interests aligned with the hack, thus making them the likely suspects. He claimed the brothers told him to tune in to WTTW on November 22, because "something big" was going to happen. The evidence seemed to be mounting, but an epilogue thread cuts the logical conclusion off at the pass with "bpoag" noting, elusively, that "J and K have been formally excluded as suspects... My original theory was incorrect."[20]

Their expungement by the online community was not based on tracking the brothers down and questioning them—that process seemed impossible—but rather on the expense of the equipment needed to pull off the heist that "simply did not exist in the wild in 1987." The finger always seemed to point toward a disgruntled broadcast industry employee testing the boundaries of exposure, dark humor, and the evasion of authorities.

VIII. Double Ghost Waves

There is something ghostlike in the nature of broadcast signal intrusions. The perpetrators become spectral media through their own legendary absence. A fracturing of identifying features occurs in the act of the signal hijacking—the reassembly of a holistic persona only becomes *more* impossible with the passage of time: voices filtered, faces covered, locations scrambled. Recovery of a face beneath a mask is obviously hopeless, and while modern technology may de-modulate a voice, the question remained: Would it be recognizable to anyone, over 30 years later?

The ghosting of the self finds an analog in the ghosting of analog television signals, too. In *Haunted Media*, author Jeffrey Sconce refers to a specific kind of "ghosting" via analog television when he documents "the eerie double-images that appear on a TV set experiencing signal interference. ... This form of interference creates faint, wispy doubles of the 'real' figures on the screen, specters who mimic their living counterparts, not so much as shadows, but as disembodied echoes seemingly from another plane or dimension."[21] With the transfer of North American satellite technology to all-digital signals, beginning in 2009,[22] the analog echo ghost has been rendered extinct. The blocky rainbow glitch and awkward frozen face distilled down into the half-inch depth of a HD flat-screen television is an efficient, if comparatively static, haunting.

Sconce cataloged narratives that verge somewhere between folklore and DIY tactics to banish the television spirit. For example, "the Travers family of Long Island, who had to turn their haunted set to the wall to avoid its scaring the children" and the Mackeys of Indianapolis, who "also received a cathode-ray visitor who refused to leave the screen. In this case, however, the apparition was not a famous entertainer, but the image of Mrs. Mackey's dead grandfather."[23]

Occasionally, we allow ourselves to be fooled. We tolerate the hypothetical wool being pulled over our eyes and stuffed into our ears. Between the intersection of design and fiction, Richard Littler's fictional Northern England town of Scarfolk is described by the author as being in perpetual lockstep with the years that preceded 1979, sharing the same general timestamp as Vrillon's Southern Television interruption. A YouTube video of an apparent broadcast signal interruption on "Scarfnada TV" (which is a parody of the regional Granada television station in North West England) begins with a test pattern arrangement in the shape of a skull, accompanied by the typical 4.5MHz sound-check blast.[24] Footage loaded with the visual cues of the 1970s places a spinning camera below a circle of people, evoking the low-angle, kinetic energy of the conception scene in the 1968 film *Rosemary's Baby*.[25] It then cuts from the people and their helium-squeak voices to an image of an eye, along with the text overlay: "WE WATCH YOU WHILE YOU SLEEP." This sequence repeats a few times. As the visible shock and sense of urgency in the photographed eye increases, the video ends with the missive: "For more information, please re-watch this broadcast." The footage, costuming, logo, and text are all a meticulous regional construction of "what television looked like" in the 1970s, branded with the hallmarks of Littler's mythical Scarfolk Council.

Why is it that the 1970s and 1980s seem an especially potent era for all things hauntological? Is it a particular despondency born of an unstable political moment, intersecting with the cultural production of artists, writers, musicians, and others in creative fields? Probably not, given that politics have fueled commentary within the arts for centuries. Is the backward-looking "retromania" idea described by author and cultural historian Simon Reynolds a factor, wherein an ever-shortening gap of time must pass before something attains the aura of the living dead, in the present?[26] Tangentially, yes. But there are a few rare examples where an event, a band, or an action appear to be persistently modern and timeless, yet immediately co-present with the past.

The subjects interviewed in Grant Gee's 2007 documentary *Joy Division*[27] point toward the long shadow of the eponymous band in this sense. Bob Dickinson's interview in this film captures one of the most apt explanations for this concept, competing with clarity against the voice of Derrida himself:

> It's about the persistence of the past in the present. And I think it's really interesting in relation to Joy Division, because, in a way, Joy Division sounded like ghosts and seemed spectral, at the time. And their music still does have that quality about it—of something that's dead, but it's alive. Something that's there and it's not there. [...] And I think in a way that they were sort of aware of that at the time that they were making those records, that the recording medium and the visual recording medium, as well as the audio-recording mediums are turning you into a dateable object and they're killing you.[28]

Mark Fisher's k-punk blog is a reliable source for explorations and applications of hauntological philosophy to popular culture. Dickinson was inspired to investigate this concept after seeing a lecture delivered by Jon Wozencroft, who was referencing Fisher, who himself was digging at the roots of Derrida's spectral writings and distilling them for a broader audience.[29] A bastardized quote from Fisher draws us back to thinking about the effect of the Max Headroom Incident, as well as Vrillon's diatribe: "Hauntology [is] *the agency of the virtual*, with the spectre understood not as anything supernatural, but as that which acts without (physically) existing ... how reverberant events in the psyche become revenants. ... The second sense of hauntology refers to that which is *already* effective in the virtual (an attractor, an anticipation shaping current behavior)."[30] This sense of "agency of the virtual" and its ability to "act without existing" was immediately experienced by audiences who witnessed the Headroom signal-jacking, and it continues to cause a vague sense of unease whenever we watch archived digital footage today. When the mundanity of everyday broadcast procedures is disrupted, it resonates at a psychic level of uneasiness.

The granular specifics of broadcast technology—much like our contemporary cocoon of wireless signals—are largely unknown among the general population. Commandeering a "mediumless medium" by harnessing unseen wavelengths results in a tailspin that comes dangerously close to clipping the history of 19th century Spiritualist communication. The voice of Vrillon emerges out of the ether, and a co-opted version of Max Headroom taunts us from a liminal space with a compressed field of view—a fiendish electronic presence accidentally channeled through an NTSC static séance. Even the "official" Max Headroom character from the television program carries a backstory; the brain and image of dying investigative reporter Edison Carter was uploaded into a computer in order to keep him alive. As "the world's only computer VJ," the "real" Headroom is a facsimile of the living—a futuristic, fictional broadcast ghost.

In their aforementioned video campaign against Scientology, the appropriation of Headroom aesthetics by Anonymous adds another layer of complexity to the mix. The collective of volatile digital spectres has become adept at harnessing the nowhere-spaces of the Internet. By destabilizing modern media, they threaten to breathe consequence back into politically and morally contentious moments of society. Air and technology pirates render themselves invisible, pure message, upsetting the medium.

Considering the paradoxical nature of these rogue copycats, which seem to float just out of step with themselves, thoughts are drawn back to the beginning of this essay, to the night drive when Ménière's disease de-coupled my own sense of hearing and reason. The mysterious combination of viral infection, inner-ear malady, and overloading of the immune system deceived my body into feeling phantom-like, if only for just a moment. This then occurred several times a day.

The acousmatic symphony of shrill tones I hear is subjective and private, but the time-shifting sense of displacement that occurs in the distance between the onset of the deep-ear, knee-buckling sea dive and the resurfacing of "normal" hearing is "dead time." During this time, I stand stock-still to wait things out while the world spins and squeals. It seems as though a lifetime of potent interactions with television sign-offs and emergency sound checks were all, in some small way, an early warning system, signaling a future need to cope and contemplate the dualistic nature of episodic hearing disturbance.

In the case of the broadcast pirate, masking—both in the literal sense of a store-bought latex mask and the metaphorical masking of the human voice with technology—creates something like a bad carbon copy. The use of the mask by the pirate was deployed for reasons of anonymity. But by obscuring one's face or voice, the phantasmatic was invoked. The deeper meaning of the word "visage" in this context captures the oscillation we feel when confronted with something that is both true and false, all the while knowing that a genuine version of a likeness is buried in the mix. In this doubling, the concreteness of a persona disassembles into a new kind of supernatural threshold—it is a bad carbon copy. If Bob Dickinson is correct in saying that "technology has turned us all into ghosts," the willing submission of the Max

Headroom pirate is an especially abject example, for he adopts the role of the grinning horror and the disturbed media spectre.

Notes

1. *When the Wind Blows.* Dir. Murakami, Jimmy T. Writer. Raymond Briggs, (New York, N.Y: IVE, 1986). Film.
2. It might have been a Devo video starring the masked and radiation-suited Boogie Boy, if you didn't know any better.
3. "WLS Channel 7—'Pirate Report' (1987)." *The Museum of Classic Chicago Television*. The Museum of Classic Chicago Television. Original air date, November 23, 1987. <www.fuzzymemories.tv/index.php?c=59&m=max%20headroom%20pirate#videoclip-2468> Accessed 29 March 2016.
4. See Wojcik, Daniel. *The End of the World as We Know It: Faith, Fatalism, and Apocalypse in America* (New York: New York University Press, 1997), 185–187: "Ashtar, who is said to be a space being and commander of thousands of space ships that together are referred to as the Ashtar Command, which will descend prior to worldly catastrophe. … Although catastrophe is imminent, believers are assured that benevolent beings are observing earth and if need be will rescue the chosen ones." (187)
5. The "Wow!" signal is an unexplained, narrowband radio signal detected by SETI member and astronomer Jerry Ehman in August 1977, while he was working at "The Big Ear" telescope at the Perkins Observatory in Delaware, Ohio. For more information, see Ehman, Jerry. "'Wow!'—A Tantalizing Candidate," in H. Paul Shuch's *Searching for Extraterrestrial Intelligence: Seti Past, Present, and Future* (Berlin: Springer, 2011), 59.
6. "WFLD Channel 32 – 'Pirate Report' (1987)" *The Museum of Classic Chicago Television*. The Museum of Classic Chicago Television. Original air date, November 23, 1987. <www.fuzzymemories.tv/index.php?c=59&m=max%20headroom%20pirate#videoclip-2465> Accessed 30 March 2016.
7. "WMAQ Channel 5 – 'Pirate Report' (1987)" *The Museum of Classic Chicago Television*. The Museum of Classic Chicago Television. Original air date, November 23, 1987. </www.fuzzymemories.tv/index.

php?c=59&m=max%20headroom%20pirate#videoclip-2466> Accessed 30 March 2016.
8. Phipps, Tony. "Can Computers Act?" *Screen Actor* Vol. 22-28 (1980), (Los Angeles, Calif: Screen Actors' Guild), 142.
9. The first use of *hacker* (in a technological, communication-oriented way) is purportedly as follows, appearing in the cover article "Telephone hackers active," by Henry Lichstein in the MIT newsletter *The Tech* (November 20, 1963): "Many telephone services have been curtailed because of so-called hackers, according to Prof. Carlton Tucker, administrator of the Institute phone system. ... The hackers have accomplished such things as tying up all the tie-lines between Harvard and MIT, or making long-distance calls by charging them to a local radar installation. One method involved connecting the PDP-1 computer to the phone system to search the lines until a dial tone, indicating an outside line, was found. ... Because of the 'hacking,' the majority of the MIT phones are 'trapped.'"
10. Volsupa. "WE RUN THIS." Online video clip. *YouTube*. Uploaded on 11 September 2008. <www.youtube.com/watch?v=j0ZFow_9vsg> Accessed 30 March 2016.
11. "Max sets bad speech habit." *Bryan Times* [Bryan, Ohio] 10 August 1987: 12.
12. Chris Knittel's "The Mystery of the Creepiest Television Hack" on *Motherboard* deserves mention here. Seemingly, we were both researching television hacking in 2013, so this year officially marks a time of rediscovery for the Headroom Incident. See <https://motherboard.vice.com/en_us/article/pgay3n/headroom-hacker> Accessed 31 October 2017.
13. See Rick Klein's online museum, *The Museum of Classic Chicago Television*. <www.fuzzymemories.tv/> Accessed 30 March 2016.
14. Hilderbrand, Lucas. *Inherent Vice: Bootleg Histories of Videotape and Copyright* (Durham: Duke University Press, 2009), 233.
15. Zoglin, Richard. "Grounding Captain Midnight: high-tech, Holmesian detective work unmasks a satellite intruder" *TIME* 4 August 1986.
16. Jennings, David. *Skinflicks: The Inside Story of the X-Rated Video Industry*. Bloomington, IN: Distributed by 1st Books Library, 2000.
17. Two of the most recent signal intrusions flipped the pornography script. In 2007, children watching Disney's *Handy Manny* cartoon on Comcast's cable feed witnessed a takeover by a hardcore pornography

film. In 2012, the same media corporation's broadcast of *Lilo & Stitch* was spliced with a new scene: six full minutes of pornography.
18. Bpoag. "I believe I know who was behind the 'Max Headroom Incident' that occurred on Chicago TV in 1987." *Reddit,* Reddit, Inc. 01 December 2010. <www.reddit.com/r/IAmA/comments/eeb6e/i_believe_i_know_who_was_behind_the_max_headroom/> Accessed 30 March 2016.
19. Phreakers (an amalgam of "freak" and "phone") are people who explore the boundaries and limitations of telephone networks. In the 1970s and 1980s, a common activity of the phreaker was to use "blue boxes" to generate electronic tones, allowing them to bypass paying for long-distance phone calls or calls at public telephone booths.
20. Bpoag. "New developments in the Max Headroom Incident mystery!" *Reddit,* Reddit, Inc. 11 October 2015. <www.reddit.com/r/IAmA/comments/eeb6e/i_believe_i_know_who_was_behind_the_max_headroom/> Accessed 30 March 2016.
21. Sconce, Jeffrey. *Haunted Media: Electronic Presence from Telegraphy to Television* (Durham, NC: Duke University Press, 2000), 124.
22. "Digital Television," Federal Communications Commission. Updated 7 December 2015. <www.fcc.gov/general/digital-television> Accessed 30 March 2016.
23. Sconce, 124.
24. Littler, Dr. R. "'We Watch You While You Sleep' TV signal intrusion 1975," *Scarfolk Council,* Richard Littler. 19 February 2014 <www.scarfolk.blogspot.com/2014_02_01_archive.html> Accessed 30 March 2016.
25. *Rosemary's Baby.* Dir. Polanski, Roman. (Paramount Pictures, St. Louis, Missouri: Swank Motion Pictures, 1968). Film.
26. See Reynolds, Simon. *Retromania: Pop Culture's Addiction to Its Own Past* (London: Faber & Faber, 2011).
27. *Joy Division.* Dir. Gee, Grant. (Madrid: Avalon, 2008). Film.
28. "Ghosts," *Joy Division.* Dir. Gee, Grant. (Madrid: Avalon, 2008). Film.
29. See Derrida, Jacques. *Specters of Marx: The State of the Debt, the Work of Mourning, and the New International.* (New York: Routledge, 1994).
30. Fisher, Mark. *Ghosts of My Life: Writings on Depression, Hauntology and Lost Futures.* (Winchester, UK: Zero Books, 2014), 18–19.

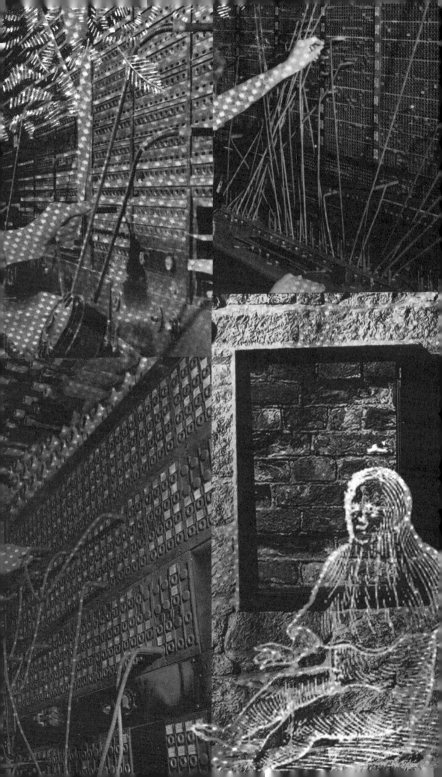

WAR, WIRE, MAGNETS

In 1900, while demonstrating his Telegraphone invention at the World Exposition of Paris, Valdemar Poulsen recorded the equivalent of the modern-day "sound bite," capturing the voice of Austria's Emperor, Franz Joseph. You can still listen to this, the oldest surviving magnetic recording, at the command of a simple Internet search. Translated from German, Franz Joseph tells the listener, "I found this invention very interesting and I thank you for its demonstration."[1] The Telegraphone, invented in 1898, was the first device capable of producing magnetic recordings.[2] Its spools held thin magnetized steel wire—this is what stored the sound—and could be used more than once to play back, erase, and re-record. The wire contained on the reels had to be thin because it took 24 inches per second to record audio.[3]

In 1903, Poulsen founded the American Telegraphone Company and began manufacturing machines for the commercial market.[4] Telegraphones were advertised as scientific instruments, as well as music recording devices, telephone answering machines, and office dictation systems. This particular machine was used by the US Navy to record the code signals of wireless transmissions—the sweeps and spaces that could lead to life-saving tactics in war. The Telegraphone was short-lived, however, because it was in competition with dictation devices like the Ediphone and Dictaphone. Ultimately, the sound fidelity of wire recordings, which had a tendency to twist and warp sound upon playback, couldn't compete with the reliability of the Ediphone's wax cylinders.

In 1928, engineer Fritz Pfleumer expanded upon Poulsen's idea for wire recordings.[5] He replaced the wire on the spools with an experiment: thin paper tape coated with iron oxide powder. Two years later, the German companies AEG and BASF licensed and then refined Pfleumer's ideas as part of a chain of improvements. This time, rather than using fragile paper tape, they deployed a new material, thin cellulose acetate tape, which was coated with magnetized iron oxide pigment and loaded onto reels. This became the K1 Magnetophon, the first practical tape recorder, which kept its reels spinning on a steady diet of BASF-produced Fe^3O^4 tape. Adolf Hitler soon discovered the benefit of time-shifting when he recorded "live broadcasts" for one city while he remained protected in another.

In 1944, Jack Mullin, a soldier in the US Signal Corp, was deployed to Farnborough, England to help solve radio interference issues. While working late, he occupied himself by listening to BBC radio—at least until they signed off at midnight every night.[6] At this point in radio history, there was, at least to Mullin's knowledge, no recordable playback media capable of capturing the quality of a live symphony. Consequently, there was no "beautiful night-music" emanating from radios in the West. The playback of phonograph discs on air was prohibited by broadcasting associations, and there were no violinists playing from melancholy studio chairs at 3am. A sweep of the radio dial after midnight would tune Mullin's overworked ears into Germany. These stations seemed to have found radio orchestras with the stamina to play music 24/7. Mullin might have rightfully assumed it was all part of a propagandized ploy to keep Teutonic wartime spirits high.

But when Mullin was dispatched to Germany in 1945 to investigate the mysteries of the country's top-secret electronics, he discovered the truth at Radio Frankfurt: The Germans had been using machines called Magnetophones. Mullin asked an official for a demonstration: "When he put the tape on the machine, I really flipped. I couldn't tell whether it was live or playback. There simply was no background noise."[7] Mullin's obsession with the absence of static had suddenly leapfrogged from a military problem-solving mission into a mission for clarity in the arena of listening for pleasure.

Mullin promptly sent two Magnetophones to the Signal Corp for study, and then "liberated" a few from bombed-out German radio stations for himself. He sensed a promising future in sound recording in the post-war era. According to military rules, anything sent home as a war souvenir had to fit into a standard mailbag. So, Mullin broke the large machines down into 35 pieces, and reassembled them once he was safe at home in San Francisco.

At a meeting for the Institute of Radio Engineers in 1946, Mullin demonstrated the Magnetophone for audio professionals from the Ampex Corporation.[8] The company was impressed, and Mullin eagerly shared his expertise in magnetic tape recording. His Frankensteined German recorders were becoming too finicky to be used practically, but their death rattles reached a small, dedicated following who believed in their power. His twin-deck Magnetophones functioned as the temporary "ear" of what would later become the Ampex 200A. While the Ampex 200A was being shopped around to radio and television stations, it was still a prototype that worked best as a playback device. Most broadcast engineers remained dubious.

Eventually, Ampex caught the attention of Bing Crosby at ABC. (Mullin had been encouraged that he might find a sympathetic audience there.) Crooning Crosby, who received the nickname "Der Bingle" from his German listeners, was tired of having to create a carbon-copy repeat of his radio performances for Eastern and Pacific time zones. When he placed an order for 20 Ampex machines at the cost of $4,000 each, he effectively financed the company's first production run, and also anticipated the widespread adoption of taped radio performances in America.[9] A very specific, circadian shift in sonic headspace occurred in the late 1940s, one that was as a direct outcome of the technical lineage that followed the path of war out of the trenches and onto wires and reels. New patterns in passive listening emerged with the adoption of the pre-recorded broadcast, which soon bled into the night. In the 1950s, insomniacs found brain fuel in the hush of tableside radios. As recording methods improved, pop-music-obsessed teenagers with transistor radios hidden under their pillows would have no memory of how things were before. While Mullin is not a household name, Bing Crosby is far from an unknown. By total chance, these two men pioneered changes in broadcasting history, albeit changes that were tied to war and wires—from Poulsen to "the old groaner," and back again.

Notes

1. "Early Tape Recordings," *Audio Engineering Society*, Audio Engineering Society, 29 December 2002. <www.aes.org> Accessed 3 August 2016.
2. "Magnetic Recording History Pictures," *Audio Engineering Society*, Audio Engineering Society, 6 July 2004. <www.aes.org> Accessed 3 August 2016.
3. See US Patent #661619.
4. Clark, Mark and Henry Nielsen. "Crossed Wires and Missing Connections: Valdemar Poulsen, The American Telegraphone Company, and the Failure to Commercialize Magnetic Recording," *The Business History Review*, The President and Fellows of Harvard College, Vol. 69, no. 1 (Spring 1995), 1–41.

5. Blundell, Stephen. *Magnetism: A Very Short Introduction* (Oxford: Oxford University Press, 2012), 88.
6. For general information about the Jack Mullin story, see The Jack Mullin Collection, *Pavek Museum,* Pavek Museum of Broadcasting, 2016. <www.pavekmuseum.org> Accessed 3 August 2016.
7. "John T. Jack Mullin," *Pavek Museum*, Pavek Museum of Broadcasting, 2016. <www.pavekmuseum.org/jmullin.html> Accessed 3 August 2016.
8. Hammer, Peter. "In Memoriam," *Journal of the Audio Engineering Society*, Audio Engineering Society, Volume 42, no. 6 (June 1994), 776–777.
9. Leslie, John and Ross Snyder. "The Early Days of Ampex Corporation," *AES Historical Committee Newsletter* (17 December 2010). <www.aes.org/aeshc/docs/company.histories/ampex/leslie_snyder_early-days-of-ampex.pdf> Accessed 10 January 2017.

WE BRIDGE THE DEAD / BREAKTHROUGH

≥ It's been far too long since I've been haunted. A friend calls to tell me they keep seeing the shadow of their dead landlord in the hallway outside of their apartment—and my reaction is jealousy.[1] I want to feel those old familiar ghost nips at my heels. The hustle up the stairs, fleeing boxes violently toppled to the floor. This occurred in the old, dark basement, where there was a secret room and a stained mattress propped against the wall. After I backed my way out of that horrible space for the first time, the hidden door would sometimes pop open to greet me while doing laundry.[2] My cats would stare into the dark yawn of the doorframe (but never enter).[3] I spent a fortune on blown-out lightbulbs in that place. Back then, I was happy to leave it behind.

Now I'm alone on the third floor of a grand Detroit library, with only the company of a man sleeping, head cradled in his arms, at the far end of the room. A pile of ghost books, the bone-crunching damp of October gloom. The smell of walnut

---

1 E421.2. *Ghosts cast no shadow.* *Fb "skygge" III 347a; *Handwb. d. Abergl. IX Nachträge 138ff; *Penzer IV 239 n. 2; U.S.: Baughman; North Carolina: Brown Collection I 682.

2 E723.7.3. *Wraith opens and closes door.* England: Baughman.

3 E436.2. *Cats crossing one's path sign of ghosts.* North Carolina: Brown Collection I 677.

shelves. Plenty of shadows.[4] One good ear full of bootlegged bass that recently arrived in an unmarked anti-static envelope. Sub-frequent headspace.[5] I want to convene with all-dead frequencies through cut-up: Raudive's *Breakthrough*, Tyrrell's *Apparitions*, decades of deep-file archive cuts. Phenomenologies of haunted looming, just behind the eyes. Willing a vision of woolen black clouds to pass over the threshold from the bubbling plaster of the cathedral halls and Echoplex stairwells behind me.[6] Summoning ghost fingers, trailing across the neck.

"I'd like to address my friends in the beyond."

Switched-on and over the radio. Layered over the muck of bass. Signal scan, fine-tuned to the open channel. Then, just static.[7] A place between stations on a medium-wave radio band—an expanse to fill in the noise. Homogenous waveforms create a beam-in site for spiritual voices to manifest. The "Breakthrough" séance is *"the possibility of projection by the unconscious onto a background of noise from radio transmitters or elsewhere."*[8]

Elsewhere, visions let loose and wander like frequencies, climbing up the roots of trees harnessed with wires to act as antennas. Too much foliage scrambles the message. Voices on tape are filtered in from the straightest tree trunk in the forest. Subjective projection and sense data gathered from the cage of

---

4 E482. *Land of shades.* Everything is done by unseen people. Type 425; Tegethoff 14; *Siuts 218ff.; Ward Catalogue of Romances II 425 (Voyage of St. Brandon); N. A. Indian: *Thompson Tales 339 n. 221.

5 E384.1. *Ghost summoned by beating drum.* England: Baughman.

6 E722.1.2. *Soul as black or white entity. Black if condemned.* Irish myth: *Cross. E481.8.4; *Dead in house of cloud.* Eskimo (Greenland): Holm 79.

7 J262. *Noisy things often empty.*

8 J2026. *Fools try to fight with man inside of drum who seems to make the noise. Are really pounding each other.* India: Thompson-Balys.

the poplar stand at the edge of a field at my childhood farm. Ouija experiments in the barn. Wire that racing planchette to the rafters for a quarter-sawn pine-beam signal boost. The lightning rod above the painted hex symbol, carrying it up to the sky.[9]

>> The residuals of Friedrich Jürgenson's magnetic tape bird calls warble in and flash me back, but from the wrong place. They form outside the window of my study, a blue-black grackle playing in the trees.[10] Camera clicks and chimes from the occasional ruin tourist in the room. Sometimes this place feels like it's more about architecture and style than words and revelation.

Another scan: "I'm trying to work with you now." And then a row of voices, a succession on the microphone. In German: "Wait!" "We are." And then back at me, directly: "Are you in salt?"[11] There is no decay in crystalline time, just mirrored black noise cycling through limitless loops. I prepare to listen deep, and listen on, to the interference.[12]

We become our own crisis apparitions. Goethe once saw his own ghost dressed in gold lace and steel-pike grey walking down a road, followed by a horse.[13] In the 1820s, H.M. Wesserman,

---

9 F51. *Sky-rope. Access to upper world by means of a rope.* BP II 511 (Gr. No. 112); Fb "reb" III 25b.—Icelandic: Gering Islensk Æventyri II 166f.; India: *Thompson-Balys.

10 F56.2. *Bird pecks hole in sky-roof to give access to upper world.* Africa (Fjort): Dennett DFLS XLI 74ff. No. 16.

11 E544.1.3. *Ghost of drowned man leaves puddle of salt water where he stands.* U.S.: *Baughman.

12 E545. *The dead speak.* Irish myth: *Cross; Icelandic: *Boberg; Jewish: *Neuman; West Indies: Flowers 431; India: Thompson-Balys.

13 E723.4.1. *Wraith returns to home and goes to bed while body is at home of friends in deep reverie.* Ireland: Baughman.

Chief Inspector of Roads in Düsseldorf, had the embarrassing habit of trying to make his spirit visible. When he became bored, he settled on wresting the apparitions of others away from themselves. Sitting in a parlor, he once rudely forced the future spectre of his female companion to appear as a widow in white. She watched her own fetch enter the room and wave a black handkerchief toward the ceiling three times before walking backwards through the wall again, wailing. This was the future to come—for her, and for him.[14]

"Wait a minute, we hear a voice again," says "Radio Peter" as he threatens to gatecrash through the static while wandering through the library, cellphone speakerphone machismo broadcasting the voice of a wounded one-night stand throughout the room.[15] The voice trails him. Then, a garble directed my way: "There are matches." A chorus of unknown voices: "We seem." And then: "Here is your Nana coming."

My grandmother swearing about the saints in swells of Quebecois French. "*Saint-Jean, maudit Criss.*" And then, her voice is clear: "Why don't you peer through the lace on the old table we used to make 'rise up' just a little bit harder?" I'm aware of how one of her eyes floats higher than the other, a family trait I didn't realize I share until recently. A washing over, the smell of crisp dryer sheets from the Laundromat where she worked in her later years, rivalling the frying chemical smell of circuit boards in the video arcade.[16] She raised me there when she was younger. "Now you need these books as portals. I would have taught you all of this." But beyond the talk of portent dreams—rotting lilies

---

14 E723.2. *Seeing one's wraith a sign that person is to die shortly.* (Cf. †F405.4.) England, U.S., Wales: *Baughman.

15 T475. *Unknown (clandestine) paramour.* Irish myth: Cross.

16 E72. *Resuscitation by smelling of moss.* N. A. Indian (Menomini): Hoffman RBAE XIV 181.

and loose yellow teeth as harbingers of death—we never had the chance to go deeper.

I'm being gently berated by the ghostly crosstalk of my grandmother. We've locked in like a laser beam. "I can see you there, when you wake up in the night."[17] It's been a problem lately. I'm thoroughly liminal from 3 until 5am. "You wander. You don't sleep." "You've done that enough... just quiet those floating feet!" For Chrissake. Her words snap back. Trashed reverb. "Without death here!" And then she is lost. But others are not.

"Can you hear us?" "You're drifting."

>>> If I could cause a haunting, I would make a cloud of snow-grey hands, bleached of life, disembodied.[18] They'd float from room to room, pointing their way from the windows to the iron gate spanning the yawning black threshold of my room. The feeling of their presence would precede their arrival in the next room. A chair scrapes beside me.[19] In George N.M. Tyrrell's theory of apparitions, the line between what we feel, think, and sense—and what we perceive—is an indistinct boundary. When I *think* a haunting, I *make* a haunting. It's a cause-and-effect collaboration of hallucinatory and material sense data. When I close my eyes, ghoul hands rush out of the room. Some say that consciousness does not rest behind the eyes.

A squelch and a hiss of spiritual kHz advance into a chopped-up whisper:

---

17 E155.4. *Person dead during day, revived at night.* India: *Thompson-Balys.

18 E422.1.11.3. *Ghost as hand or hands.* England, U.S.: *Baughman.

19 E539.4.1. *Ghostly chair in cellar jumps up and down on three legs, points with fourth at spot on floor. Witnesses dig up body from under floor.* U.S.: Baughman.

"The smoke spoils." ... "You have no helpers." [...] "The trick is in the flame."

I recently encountered a lightbulb that passed through the hands of Sir William Crookes, the famed scientist and discoverer of plasma, as well as the channeler of cathode rays via vacuum tubes, and a victim of Spiritualist shenanigans. I'd hoped to feel something when I held the Crookes bulb, but it just felt like a lamp with loose wires.[20] The materialized spirit "Katie King"—a controversial conjuration made by the teenage medium Florence Cook—sometimes sat on Crookes's lap during séances. King's apparitional form moved around the darkened room, bestowing wispy eyelash-kisses and holding hands with those sitting at the séance table.[21] How do you arrive at third base with a spirit? Crookes and King (or was it Cook?) had no option but to continue their ruse. He was in love with a living ghost.

The eavesdroppers of thoughts respond:

"Linger here, love." "I cannot sing for you."
"Don't tire yourself. My hair has been cut off."
"Raudive, change the subject." "I hear you." "That's right."

---

[20] E765.1. *Life bound up with light (flame)*. Breton: Sébillot Incidents s.v. "vie"; Fb "lys" II 483ab; Gaster Thespis 275f.; Icelandic: *Boberg.

[21] E471. *Ghost kisses living person*. *Fb "gjenganger" I 444a.

V. PLAYING THE SPECTRE

HUMMING THROUGH SALT
AUDINT'S Spurious Frequencies

I. *Weight, Calm*

A slow dive into the world of AUDINT is a little like the old folklore fallacy of salting the tail of a bird. Or, even better, its variation: salting the backs of witches. The analogy holds that through the impossible task of sprinkling salt on either bird or witch, one gains power over it—slows it down, weighs down its insubstantial or contested body. To pour salt on the backs of AUDINT's researchers is to give weight to the metaphorical body of their archive of ghosted histories. To slow down the algorithm—a profuse generation of texts and tendrils and esoteric wanderings and viral intrusions—so that we might find the space needed to submit to the idea that truth is stranger than fiction, and that the AUDINT divide tends to err on the side of truth. We could run our rattrap straight into their rabbit hole and waste days doing online research into people that couldn't possibly have existed, or into government programs that sound fake but turn out to be real. The jarring effect of this quickly becomes disturbing. We could willingly submit to virulent websites and translate German in desperation, to seek out the biographies of sketchy historical figures. I won't lie: I did all of this. In the end, chucking an entire salt shaker in AUDINT's direction might have a better chance of slowing the RPMs enough to make the connection, than my cautious castings of singular grains.

Formed in 1945, AUDINT was originally composed of Hypolite Morton,[1] Bill Arnett, Walter Slepian, and the ill-fated Eduard Schüller. Since its 2008 incarnation, Toby Heys and Steve Goodman have been two mainstays, with others floating in and out. AUDINT claims its original members had formerly been with the Ghost Army, a mobile WWII deception unit populated by artists and theater technicians. Famously using "battle DJs" who remixed reality with field recordings of the sonic imprints of war—and using inflatable tanks as props—the Ghost Army was a diversion tactic designed to confuse Nazis on the whereabouts of Allied troops.

An encyclopedia of acronyms and code names form the brick and mortar of AUDINT's foundation. They consider themselves a "splinter research cell" with nestled connections to similar cells: OSS, Operation Paperclip, OAR, JIOA, Operation Wandering Soul, SAIN, and others, all accounted for in the system of files known as the *Dead Record Office*. AUDINT itself is a contraction of "audio" and "intelligence." The arc of the whole outfit encompasses a subliminal world, one where the institutional language of organizations has been amplified into something unsane, unsound, a network where rogue operatives left to avoid stepping on the mortar-blast eggshells they have laid out for themselves.

My own Canadian uncle, Orvy Butler, was captured in Italy during Operation Husky, the code name for the Allied invasion of Sicily in 1943. His escape was partially thwarted by the weight of the field radio he carried; he lingered too long, and was sent to a German POW camp, where he was held for three years. There, he was fed a diet of stewed mystery meat and was kept in a half-flooded room where limbs were prone to become ghost limbs in the damp. Back home, the Canadian Red Cross declared him "presumed dead." His wife moved on and had another child. After American troops liberated the camp, he eventually showed up at her door one day, as a former ghost of himself, trailing smoke and lost signals.

AUDINT was born out of a crisis, a reification or an embodiment of the damaged psychic landscape of the years following WWII and its veiled, surface-level "schizo-lithium calm."[2] It was initially founded out of a sense of cooperation between militaristic powers. But, having succeeded in its assigned directives a little too well, its members defected and spirited away their research to the underground, "that space in which technological innovation meets speculative thinking…"[3]

II. *Salt*

The presence of AUDINT in the Windsor-Detroit area was a subterranean infiltration. With salt. And with a most aberrant effect of sound. The Detroit River splits the two cities apart on the surface. But 1,200 feet below the surface, a massive 1,500-acre salt mine connects the two countries.[4] A hundred miles of subterranean road loops itself into knots, in the darkness. The mines exist as a network of rooms held up by salt pillars—solipsistic architecture that is a result of the blasting method. The interior walls are sheared off with explosive charges. The collapsed chamber is hollowed out again, crushed and driven up by a conveyor. There, the shock of surface air hits deposits that haven't seen the light of day since the Devonian period, 400 million years ago.[5] The mine blasts decouple the slow capital production of the earth.[6] Graveyards for historic heavy machinery and the bony husks of horses are said to be down there too—labor not wasted on the disassembly to haul them back up to the surface. Locals remember riding down into the mines for public tours, past the Saint Barbara shrine—the miner's saint of safety—and spilling out with schoolmates into the sparkling rooms. Licking at the air.

In the late 1700s, Ernst Chladni, a pioneer of acoustic research, visually exposed the hidden properties of sound, which served to provide a template for instrument-builders to shape their guitars and violins in order to maximize the richness of resonance.[7] In the classic Chladni plate experiment, grains of salt are dusted onto metal plates. When the plate edge is run over with a violin bow, the vibration causes the grains to reorganize into corresponding patterns of resonance. The patterns are formed by the plate's nodal lines—areas where the surface stays still. Those blank spaces *between* sound are where the salt attracts, creating Chladni figures. Those metaphorical nodal lines—the spaces of negation and collision—are where AUDINT's sonic research interests lay.

Salt is contradictive. Tossed off over the shoulder it is a cleansing—or a superstitious tic. At a chemical compound level, the sodium pentothal that plays a part in AUDINT's animated film, *Delusions of the Living Dead*, gains sentience as an agent of truth. But for those who fall under its power—narcosynthetes—their revelations pass into the territory of the legally inadmissible. Surrounding an unclean house, it is protective. And yet the mining of salt in the Windsor-Detroit area once generated legends about the inevitable collapse of the region down into its own bowels. A sinkhole *did* appear at the Windsor side of the mine in the 1950s, around the same time AUDINT was blasting away at the degenerating edges of their colleague Eduard Schüller's mind.

III. Cipher

In the fall of 2015, AUDINT's *Delphic Panaceas* exhibition was on view at Artcite in Windsor, Ontario. Their *Martial Hauntology* project is a cipher to the expanded works in the exhibition—a collation of four years of sonic research, released in a limited edition of 256 copies. An LP contains two chapters: *Delusions of the Living Dead* and *DRNE Cartography*. A selection of cards from the *Dead Record Archive* and a 112-page book, *Dead Record Office*, record the history of AUDINT.

With the *Dead Record Office*, AUDINT have created a historical record of the acoustic and unsound universe. The archival information cards hint at the presence and infiltration of AUDINT's agents throughout the history of physical sound media. Innocent devices are hijacked electronic witnesses to experimental agendas. The media formats that they feed upon record, reassemble, and broadcast the sonified fodder of PSYOP, and for the *Ohrwurm*, which we have all experienced.[8]

Histories of technology are filled with hypothetical worlds running in tandem with unauthorized adaptations overtaking intended use. For every Edison phonograph, there are consequent mumblings about rumors for lost plans for "ghost telephones." Historical figures like the Danish engineer Valdemar Poulsen are recognized as essential conduits in the growth of affective sonic experiences in the 20th and 21st centuries. And for every house cat we cohabitate with, there is the paranoia that we could be living with an "acoustic kitty," a vile and delusional program launched by the CIA to send "trained cats" equipped with surgically implanted antennae and microphones to spy on Russian embassies. Did anyone in the CIA ever *really live* with a cat?

The animated video installation *Delusions of the Living Dead* is "the story of sound recordist and designer Walter Slepian's

1949 plan to purloin and photograph French neurologist Jules Cotard's notebook, an arcane medical document pertaining to the process and methods required to 'seed walking corpse syndrome into a subject's bed of cognition.'"[9] The title of the show, *Delphic Panaceas*, is in fact a reference embedded in the same film. According to Toby Heys, it's a "black humored reference to the idea that Cotard Syndrome could be an answer to AUDINT's problems of connecting with the undead."[10] The soundtrack integrates the sonically audible voice of "Ms. Haptic" against soundscapes composed by Goodman and Heys, as well as the more insidious use of an integrated infrasonic band, hidden among the digital weeds.

IV. *Hum*

One day, as I sat at the breakfast table, skimming through a pile of newspapers that should have been recycled months before, I felt the vague disturbance of familiarity. As the smoke cleared, I realized the photograph on the front page of the *Detroit Free Press* was of my old apartment in Windsor, Ontario. Apparently, the sickly yellow stucco walkup in Sandwich Town had become newsworthy as a victim of "The Windsor Hum." Reports of Hum-like activity are global, reaching back to the 1830s before establishing a push-pin presence across maps throughout the 1970s. Wherever it appears, etymology melds with geolocation: the Taos Hum, the Bristol Hum, the Auckland Hum. In early 2011, Windsor developed its very own Hum, a mysterious infrasonic event spiking deep at 35Hz. When residents woke late one night to a low-frequency rumble, slashing open their curtains to yell at an idling car booming bass, they instead found empty, dark streets.

In 2002, while standing, insomniac-prone, in that same Windsor walkup, I looked out the kitchen window to find the sky on fire—a total apocalyptic vision over Detroit. Gigantic orange plumes trailing up into a gradient of anemic ochre, wretched green, and hazy purple. A low-grade thrum was prickling at the soles of my naked feet. Assuming some great industrial disaster was about to roll toxic fumes over the river, I pounded on my roommate's door and stuttered about the cataclysm in the sky. This is when I first heard the (pillow-groaned) words "Don't worry, it's just Zzzzzug Island." (More on Zug Island in a moment.)

Describe the Windsor Hum. A deep-time bass rattle. A quivering in the gut. The creak of double-glazed windows with an angry bee caught between two planes. Night terrors. Not everyone can "hear" the Hum, but the vibroacoustic effects of infrasound—sound that exists below the range of human hearing—can cause suffering, insomnia, depression, anxiety, and migraines. Most victims of the Hum describe it as something *felt* more than *heard*. Their bootlegged bodies suffer incessant monotone pressure, beating on their eardrums. Rational finger-pointing toward local heavy industry was counterbalanced with viral conspiracies: trending UFO reports, ionospheric interventions by HAARP (an ionospheric government research program), and flyovers by experimental military aircraft.

Both the Hum and infrasound can mimic the tropes of a traditional haunting. In the early 1980s, scientist Vic Tandy was likely surprised to find himself collaborating with psychical researchers; together they traced the cause of a recent "haunting" outbreak in his laboratory to the installation of a ventilation fan. That recent cold-sweat feeling of dread and the shadowy apparitions stuck in the corner of Tandy's eye were linked to inaudible infrasound being produced by the fan, which operated at a steady 18.9Hz. Tandy's vibrating eyeballs were allayed by the removal of the fan.[11]

The Hum is a more holistic environmental phenomenon,

one that defies easy solutions. It is a fake-out haunting, a physiological response to the invisible effects of the discord between environment and industry. It's an ominous protest of the Anthropocene in the form of infrasonic terror. Salt and steel dance on air, down into the lungs.

As for the Windsor Hum, attempts to trace its location (nevermind its cause) have been an exercise in frustration. Like describing neural pain or ghost limbs, pinpointing where one flesh ends and the other ghost twin begins, the Hum's oppression seems to come from everywhere and nowhere. First, the semi-trucks idling on the crumbling Ambassador Bridge joining Windsor and Detroit were blamed. But this theory belly-flopped into the river below.

In 2013, hard-noised scientists finally captured the "temporal and spectral" signature of Windsor's Hum, describing the process as being "like chasing a ghost."[12] Fingers pointed toward Zug Island transformed into tenuous high-fives. The electric arc blast furnace at the US Steel plant was hemorrhaging infrasound and VLF waves across the border. These waves have been identified as the "likely" cause of the Windsor Hum and are believed to cause the elusive soundtrack of the aurora borealis. On Zug Island, the resistance of steel being magnetized back into its base elements reverberates like the wailing of entangled souls, while offsite it bleeds over the border, damped down into a low-pressure menace. The Hum continues to beat the ears of the city in unpredictable fits of biomechanical violence. The *Ohrwurm* of industry is always there.

V. *Clash*

There are natural and celestial events that predate the explosive violence of either of the two world wars. The Chelyabinsk meteor impact of 2013 generated some of the largest global measurements of infrasound to date (with a shockwave strong enough to shatter windows). If the measurement of such things had existed at the time, the 1908 Tunguska meteor would have been off the charts. When Leonid Kulik, curator of meteorites at the Mineralogical Museum in St. Petersburg, visited the Tunguska impact crater 19 years later, locals were still reluctant to talk about or visit the site. There was homegrown folklore that characterized the explosion as a curse from Ogdy, the god of thunder and infra-bass, who smashed the forests and chased off animals as a punishment for some uncited human slight toward the deity.[13]

At the core of AUDINT's *Dead Record Office*, there are a few essential moments: the development of the TwoRing Table, and the entwined fate this device shared with Eduard Schüller, the fourth member of the first wave of AUDINT. Schüller was a German expat audio engineer who moved to the US after WWII. While working for AEG during the war, he observed "how the dashing SS officers interfered with [technical] development, and how Jews suddenly went missing."[14] Schüller (1906–1976) was a communications pioneer, working for the Hertz Institute in 1932. Soon after, he was at AEG. Schüller was central to the development of the magnetic tape recorder, working with Fritz Pfleumer, inventor of paper tape recording. He collaborated with BASF to create magnetized medium on plastic tape. His patent—"magnetizing head for longitudinal magnetization of magnetic recording media"—was granted in 1933.[15]

The AEG Magnetophone (1935) was a heavy, yet portable precursor to the mobile reel-to-reel. American soldiers encountering these systems dismantled and fragmented them, shipping them piecemeal back to America for plunder. Schüller was the weak link in the group, due to the black mark of his Germanic ancestry (despite his lack of connection with Nazi tenets). When a test subject was needed for AUDINT's phono-experiments, Schüller volunteered. His alienation was sealed after Arnett and Slepian met with Alan Turing, who planted the idea for the device they would later name for him, out of reverence: the TwoRing Table.

With the TwoRing Table, the themes of the collision and the fragment are key. According to the group's mission statement, "It is the collision, in all its vibratory formats and excesses that interests AUDINT."[16] Like a mutated turntable, the TwoRing spun a locked-groove record. As the two individual arms of the TwoRing Table were propelled backward and forward, a specific sonic collision occurred once the arms met: the soundclash. This clash—and "the waveformed artificial intelligence" at its core— arrived, "meld[ing] matter with anti-matter, and fuse[d] the covert back-masked message with the overt narrative refrain."[17] In this collision of three sound types, normative reality was displaced, and a certain kind of threat rose up to take its place.

AUDINT's investigative history, from this date onward, proposed to find ways to "open the 3rd ear," based on the simultaneous deployment of audible sound and the channeling of ultra- and infrasound—sound waves that exist above and below human hearing. Ultrasound exists above the upper registers of human hearing of 20kHz, but is recognizable in the shriek of a (silent to us) dog whistle or the echolocation of bats. Infrasound dips beneath the 20Hz range where "typical" human hearing bottoms out. Infrasonic specialists read the presence of these tones in tracking earthquakes, volcanic eruptions, and nuclear and aerial blasts, as well as the geological rearrangement of rock (not music) underground. Animals running away from a forest serve as an early warning system, a clear manifestation of their sensitivity to lower frequencies. None of these things are good.

AUDINT took the potential of the TwoRing Table one step further by adding *three* turntables to the mix. They played "GITH repeater" discs on the decks to form a sound-based triangulation, where the heterodyned intelligence discharged "sonic shrapnel known as hooks into Schüller's neural flesh."[18] His sanity began to stretch and unravel. The hook created an earworm, or *Ohrwurm*, which awakened Schüller's 3rd ear. In short, the *Ohrwurm* caused Schüller's consciousness to mutate. And, with a decoupled 3rd ear, his corporeal flesh began to take on the physical appearance of the media that it absorbed. His skin was, essentially, growing grooves. Wrapped in reels of magnetic tape, Schüller was turned into a "monstrous payload," an "embodiment of excessive communication."[19]

The body is never silent as a medium. In the earliest versions of the 1940 anechoic chamber at Bell Laboratories, known as "the world's most silent room," the sounds of one's own mortality are amplified and broadcast back unto themselves—the internal sounds of the ears, the electric hum of the nervous system. It's an unsettling experience; sanity dissolves quickly, and it has been claimed that no one has managed to exist in this desensitized mode for longer than 45 minutes.[20] A memory of a sound bite from *Delusions of the Living Dead*: "All sensory information is spectral in essence."

VI. *Crystallization*

AUDINT leans toward materiality often, letting loose with talk of the "alchemical properties of shellac."[21] In popular knowledge 78RPM records were collected during WWII "record drives." Their shellac was extracted and repressed into new records, or used as wire insulation in field radios and military vehicles. Music was seen as disposable, and access to new recordings of popular music could only come from the literal (and physical) negation of the old. In their founding years, AUDINT discussed "the implications of melting down records and casting them into forms, which can be used for ritualistic practices."[22]

Sonic research is self-referential, pulled from the "vast sound library of the atmosphere."[23] In order to unlock the sound that exists on material bases (be it a disc, a tape, a hard drive, or a cloud), it must be commingled into a new network, one where signals move through the geologic (germanium), filtered electrical impulse, magnetic, circuits, speakers, what have you. While theories of residual hauntings linked to geophysical deposits of limestone are speculative guesswork, the communicative properties of quartz crystals are hard and fast. Quartz responds with an electrical backlash when placed under pressure. The skating of a phonograph stylus through a recorded groove puts stress on the quartz tip of the pickup. An electrical response acts as the signal that becomes the message to be heard.

Quartz stabilized radio frequencies. It was also the backbone of crystal radio sets, sonography used to locate artillery in the WWI era. All of these devices worked on mineral frequency. The crude methods of growing crystals were not up to mass-production snuff, and so Brazil was stripped of a portion of its natural quartz crystal.[24] The 1943 Reeves Sound Labs film *Crystals Go To War*[25] anthropomorphizes the synthetic rocks grown in labs, filtering the voices of soldiers bouncing back from aircraft above or entrenched with radios in hand below. In the presence of an excess of vibration, however, these same crystals that provided constancy in sound could also recoil and begin producing "spurious frequencies."

In 1917, while working for Bell Laboratories, Alexander Nicholson turned to Rochelle salt when building one of the first crystal-controlled oscillators.[26] All that Detroit rock salt (*sodium chloride*) being hauled up from the mines was inferior in its singularity. Rochelle salt (*potassium sodium tartrate*) is known as a "double salt" with piezoelectric properties. Double salt, like the TwoRing Table or Magdalena Parker's ring modulated loops (discussed below), is tangentially connected with the "3rd ear" concept so often cited by AUDINT. As a mixture of two simple salts, polluting one another's purity, a new crystalline structure is created, divergent from the two original structures. Most importantly, the electronic devices that powered the early years of AUDINT's research—microphones, radios, headphones, recorders—were operating on crystal power and salt derivatives. In the end, all those solid Rochelle crystals were prone to melting back into air.

VII. *Oscillation*

In the *Delphic Panaceas* installation, a set of headphones plays the aural archive of Magdalena Parker, a second-wave member of AUDINT, recruited in 1959. She was a "Chilean performance artist and filmmaker whose experimental work revolves around the use of the voice, magnetic tape, and the production of sonic cutup collages for ritual incantation,"[27] according to author Patrick R.J. Brown. The headphones in the installation repeat the contents of a compact cassette tape containing 15 analog synthesizer and tape-loop sequences. It is joined by Side C of the project, a 16th loop encased in resin, "to ensure that it is never heard,"[28] as Parker instructed, in a move that hints at the dualistic properties of banishment and protection. The loops were created by way of ring modulation—a type of electronic sound processing that, like the TwoRing Table, relies on the multiplied collision of two separate signals. The subsequent sound is both the *sum* and the *difference* of each individual waveform. Both signals are layered on top of one another, but there's also a *new* signal that expresses the *differences between* those sounds—a mutant third layer.

You could say this effect is another example of AUDINT's 3rd ear Other. If ring modulated frequencies are not in harmony with one another, the final sound is one of metallic bell chimes. The individual loops in Parker's aural archive are constructed of those woozy, repetitive chiming frequencies. They could easily induce the listener into a trance state, if it weren't for the safety valve of fragmentation between sides, as dictated by the cassette-tape format. It's somewhat appropriate, given Parker's apparent ability to induce states of possession among audiences, and AUDINT's own descent into the psychic history of the clang of the bell at the New York Stock Exchange, or the cross-dimensional communication potential of Vietnamese gongs.

VIII. *Queue*

As AUDINT's second wave of members begin to bridge the analog-digital divide, new kinds of spectral economies arise. They tumble into the zone of reified technology, instigated by economic trade via IREX (irrational exuberance) and the waking of a sentient, supernaturalized virus called IREX2. Lying dormant in the hollows between those Chlandi plate nodes, IREX2 is pragmatic and self-replicating, awaiting the growth of the networked tangle of the World Wide Web to become thick enough so that it can run rampant without detection. IREX2 "breached and convened the rules of the spectral masses by converging the human, the algorithm, and the unearthly into a singular audio intelligence."[29] It was apparently IREX2 that led to AUDINT's current ranks being occupied by Toby Heys and Steve Goodman, who were recruited/coerced into becoming members by "a spam email and Trojan horse" in 2008, due to their backgrounds as sonic researchers and artists, as well as their "sketchy purchases" made on the Silk Road.[30]

There is a certain kind of danger implied in this kind of artistic research—a non-specific paranoia for the witness who falls too willingly and fully into AUDINT's altered reality. Much like the work of Trevor Paglen,[31] who hunts, and confirms the existence of "programs that don't exist," along with government and military apparatuses, and photographs them using long-range lenses, the "innocence" of artistic research suffers under the nefarious threat of becoming invaded by wild and conspiratorial forces. This is something that one imagines resulting in a threat, or a disappearance, more so than a cease-and-desist order or NEA-era public-shaming tactic.

I write about these elusive ideas of salt, of sound, of the materiality of media formats and landscapes, of the minutiae of the geological, of local history, in order to position AUDINT into the regional background in which their exhibition appears. Realizing the power they'd created, the collective realized the need to destroy the soundclash-producing GITH discs that operated via the TwoRing Table. The systematic fragmentation of these discs, shattered and split, were pollinated in plain sight and scattered into mainstream listening culture. Sound bites were dispersed onto sound effects, stereo fidelity, and test-tone recordings, which isn't to imply that the reassembly of the whole will ever occur, it's more that nothing ever really disappears. Sound is a social beast, hungry for circulation.

Notes

1. An interesting link to the name of a salt company—Morton Salt—and the name of a road in Windsor that serves as an entry to a salt mine.
2. AUDINT. *Dead Record Office* (New York: AUDINT Records 2014), 6.
3. AUDINT, 86.
4. "History of the Detroit Salt Mine," *Detroit Salt Co.* <www.detroitsalt.com/history/> Accessed 28 January 2016.
5. Ibid.
6. In 1906, a 1,000-foot shaft was pierced into the heart of what would become the underground rock salt city. Now that salt is used to de-ice

Michigan roads. It eats away at crumbling concrete and melts the city and its cars season by season.

7. Whether superstition or reality, the act of submerging drum cymbals into saltwater baths ostensibly opens the metallic pores and perverts the sonic quality, making them ring more complex and "dirty," with a more-subtle attack. Aesthetically, oxidization tampers with manufacturers' finishes, removing the sheen of the commercial and allowing the musician to partake in a sort of "haunted now," with a cymbal set that looks antiquated, inherited. But this is a false inheritance—it is one that is bought and superficially brought into the hauntological now.

8. The "Ohrwurm," or earworm, refers to the experience of having an infectious piece of music stuck in your head. Sometimes the earworm manifests with no known prompt (a song randomly pops into your head). Or, after hearing a piece of "sticky music," the sounds can run through your mind for days.

9. "Sonic Disruptions," *Tate Modern*. <www.tate.org.uk/whats-on/tate-britain/talks-and-lectures/sonic-disruptions> Accessed 28 January 2016.

10. Heys, Toby. "Re: Spurious Frequencies." Message to Kristen Gallerneaux. 1 October 2015. Email.

11. See Tandy, Vic & Tony R. Lawrence. "The Ghost in the Machine," *Journal of the Society for Psychical Research* 62, no. 851 (April 1998): 360–364.

12. Novak, Colin. "Summary of the Windsor Hum Study Results," *Global Affairs Canada*, Government of Canada, 23 May 2014. <www.international.gc.ca> Accessed 8 January 2017.

13. Phillips, Tony. "The Tunguska Impact—100 Years Later," NASA Science: *Science News*, NASA, 2008. <www.science.nasa.gov/science-news/science-at-nasa/2008/30jun_tunguska/> Accessed 28 January 2016.

14. Translated from the German, a biographical entry at the City Museum of Wedel.

15. German patent, DRP 660,377.

16. AUDINT, 17.

17. AUDINT, 18.

18. AUDINT, 19.

19. AUDINT, 26.

20. Eveleth, Rose. "Earth's Quietest Place Will Drive You Crazy in

45 Minutes," *Smithsonian Magazine*, December 17, 2013. <www.smithsonianmag.com/smart-news/earths-quietest-place-will-drive-you-crazy-in-45-minutes-180948160/?no-ist> Accessed 23 January 2016.
21. AUDINT, 17.
22. AUDINT, 17.
23. "Delphic Panaceas: An Infrasound / Ultrasound and Video Installation," *Artcite*. <www.artcite.ca/history/2015.html> Accessed 28 January 2016.
24. After the war, Bell Labs refined a viable process. By 1970, virtually all crystals embedded in electronics were grown in labs. But those older crystals, naturally derived, do they have the potential to tap in to deeper landscapes?
25. *Crystals go to War.* Reeves Sound Lab, 1943. Prelinger Archives. <www.archive.org/details/prelinger> Accessed 28 January 2016.
26. Brown, Patrick R.J. "The Influence of Amateur Radio on the Development of the Commercial Market for Quartz Piezoelectric Resonators in the United States," *Proceedings of the IEEE Frequency Control Symposium* (1996), 58–65.
27. AUDINT, 33.
28. "Magdalena Parker," AUDINT. <www.audint.bigcartel.com/product/magdalena-parker> Accessed 28 January 2016.
29. AUDINT, 84.
30. Heys, Toby. "Re: Spurious Frequencies." Message to Kristen Gallerneaux. 1 October 2015. Email.
31. See Paglen, Trevor, and Rebecca Solnit. *Invisible: Covert Operations and Classified Landscapes*. New York, N.Y: Aperture, 2010.

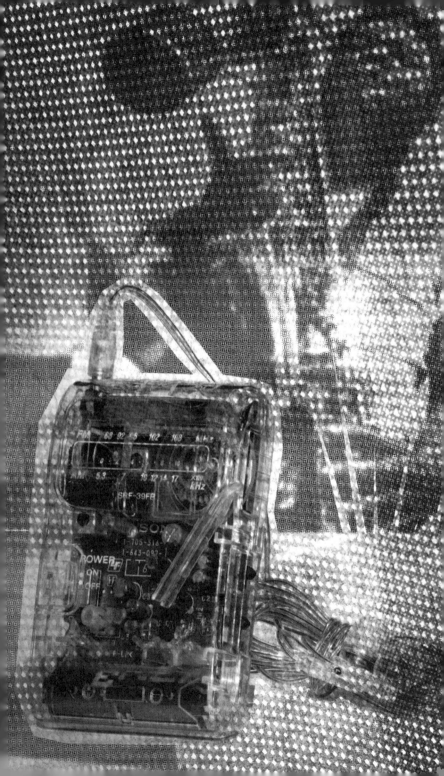

PRISON RADIO / SHOTGUN

One afternoon, a small, slightly crushed cardboard box appeared on my office desk at the museum. After slicing through the packing tape, a stench cloud of cigarette smoke plumed upward and smacked me, headlong, in all of the senses. I had a mental image of the person who packed it sticking a straw into a corner of the box and exhaling an entire cigarette inside before sealing it. The almost-hilarious intensity of the smell had my eyes watering. When another curator walked by, he quizzically sniffed the air before following his nose toward my office door. Poking his head in with a concerned look, as though I'd suddenly reached some critical stress level and just hauled off and lit up a cigarette, I guiltily verified, "I'm not smoking in here, I swear!"

Pushing aside the top layer of Styrofoam peanuts, I found a transparent Walkman camouflaged in a nest of newspaper. When I gingerly picked it up, a blast of sound struck like a raptor. Jr. Walker & the All Stars' "Shotgun" spit out of the headphones at full volume, like some kind of Motown defense mechanism. I must have hit the power switch.

The Sony SRF-39FP is one of the most popular portable AM/FM audio players used in prisons across the United States. Sony marketed its first SRF-39FP (the "FP" stands for "Federal Prison") over 15 years ago, and have kept it in production ever since. The radio's defining feature is its transparent plastic body and clear headphones, designed to prevent the storage of contraband by inmates. Interior transistors, resistors, and circuit paths are laid bare. These radios are only available through approved prison suppliers, who provide many other kinds of clear-cased personal effects, designed for sale in prison commissaries. The convenience of the radio is twofold. Not only does it provide inmates with private and personal agency, it also makes routine cellblock checks by correctional officers more efficient.

This specific type of transparent industrial design first found its way into consumer electronics with the refinement of materials like Lucite and acrylic. One of the most common devices to undergo this treatment was the telephone, the shell of which was transformed into crystalline windows that encased an interior architecture made up of discreet components. Though the true origins of the first clear telephone have proven elusive, they began to appear irregularly in the 1960s, often as true prototypes rather than saleable products, and then began to flood the market in the late 1980s through the mid-1990s. Many of these phones found their way into the bedrooms of teenagers (this author included). Bell South, Conair, RadioShack, Swatch, Unisonic, and others capitalized on the trend, installing more whiz-bang features like visible bells, flashing lights to indicate incoming calls, and glowing blue neon tubing.

Other devices and personal accessories received similar treatment: boom boxes, televisions, backpacks, raincoats. In an era when owning a home computer meant living with sickly beige plastic things, Apple Inc. made several products that broke the mold of bland ubiquity. The original translucent white and Bondi Blue iMac G3 was released in 1998. While it was more guarded in its transparency than a Unisonic telephone, it refreshed sad computer dens on a massive scale. They soon offered extended palettes of tangerine orange, lime green, raspberry, grape, and blueberry.

These mechanized and miniaturized worlds of the transistor and silicon age were visually amplified and coded with the synthetic fruit-flavored palette of the late 20th century. The crystal clearness of the telephone amplified how little we know about basic communication channels, and the backlit glow of the iMac hinted at worlds of possibility that were *almost* visible. In both cases, these design cues were not meant to educate, but to further expose the mystical properties of complex technology. It's almost as though clarity confirms our ignorance.

The Sony SRF-39FP radio is simple but powerful, as evidenced by the clarity of the WJR signal that ran out through the earbuds. I even found some of the more obscure college radio stations with ease. The SRF-39FP's copper and ferrite antenna has the power to reach far, and its tuning responsiveness is robust. Powered by a single AA battery, it can lend over 40 hours of aural escape. Battery efficiency becomes important in prison, where inmates are only allowed to spend about $300 a month on commissary items. Less money spent on batteries means more money for toiletries and snacks.

Radios like the SRF-39FP continue to reign over digital players and MP3 players for good reason: The cost of an iPod is at least three times as much as an AM/FM radio. An iPod also requires a further investment in purchasing music, and the MP3 playlists available to inmates are highly censored, stripped of explicit lyrics and content that could potentially incite violent behavior.

Radio has historically acted as an outreach technology, where cultural and political developments can be received (and, in its early days, transmitted) from the comfort of home. The Sony SRF-39FP is at odds with itself, invisible among the general public, yet widely adopted within the internal community of the Federal Prison System. Owning a radio in prison allows inmates to reestablish a sense of privacy, comfort, and selfhood via an escape into the comforts of music, entertainment, news, and weather updates. And this radio, simultaneously a translucent conduit and a tool for making unseen/unheard people more substantial, enables inmates to tap into wider communication networks—to reinsert themselves into a dialogue with the outside world.

It has other uses too. Televisions in common areas in prison operate similarly to drive-in movie theaters—a low-powered transmitter provides a radio frequency to tune in to, by way of an external device. In this way, prisoners can tune their radios to receive the sounds of television, encasing themselves in the comfort of a sound carrier bubble, delivered as a mainline through headphones.

Despite the fact that it's illegal for inmates to transfer personal property amongst themselves, it is a tradition for outbound people to gift their radio to an inmate left behind upon their release. Many former inmates believe that keeping the radio after their release is bad luck—a physical manifestation of bad times and bad memories, or a breach in the informal "convict code," which dictates its own complex systems of exchange. Owing to this, the cycle of circulation makes it somewhat difficult to attain radios. With the establishment of collecting communities powered through digital marketplaces like eBay, however, these radios occasionally come up for sale. Evidence that the radio sitting on my desk once belonged to an inmate is concrete: Etched into the plastic at the top of the radio is the former prisoner's ID. Despite attempts to contact the seller about further information, none was offered.

BREAK IN TAP

PERFORMANCE RITUALS
The Moog Prototype

When the experimental composer Herbert Deutsch first met synthesizer pioneer Bob Moog, he told him he wanted an instrument that didn't exist. He wanted something that could "make these sounds that go *wooo-wooo-ah-woo-woo*."[1] At this same music educator's conference,[2] Deutsch first diverted Moog's attention away from the Theremin kits he was selling by asking "Do you know anything about electronic music?"[3] Six months later, in the summer of 1964, Deutsch arrived in Trumansburg, NY with grant money in hand from Hofstra University, where he taught. He and his family set up camp at a nearby state park; this was to be a working vacation. Moog greeted Deutsch at his front door with the declaration: "We're gonna work in my studio... which is in the cellar."[4]

A few months prior, Moog's musical worldview was expanded at a Greenwich Village studio, where he watched his new friend perform an experimental concert on the artist Jason Seley's bumper sculptures, turning car parts into impromptu instruments with his percussive mallets. Deutsch was an early entrant into the world of tape manipulation, too, forming sounds replayed from recorders into spliced symphonies created from patched adhesive and miles of polyester spindled onto reels.

For two weeks that summer in Trumansburg, the pair shot ideas back and forth. Moog's electrical engineering skills and collaborative spirit complimented Deutsch's experimental, musical talents. At first, they didn't even call the thing they were developing a synthesizer, but rather "a portable electronic music studio."[5] Exchanges over whether or not to give their instrument a keyboard, the qualities of attack and decay, and one spectacular epiphany about a doorbell all coalesced to give form to the instrument being birthed, Frankenstein-style, in Moog's underground workshop. The short version of the story is that Deutsch soon began to hear the first signs of his electronic "wooo" and "ah" sounds escape from the cellar. It is rare that the origins of sonic specificity can be traced to an exact moment, but the birth of the structure of "the Moog sound" (and by extension, a new way of making electronic music) is well documented.

A key moment demands a digression. One day, Moog sent Deutsch to retrieve a doorbell button from the local hardware store:

> ...within a half an hour or an hour at the most, he had set up the idea that if I press a key down on this little keyboard, it would allow the sound from an oscillator to play. He said, "When you [press a key], press the doorbell button at the same time." And what he had designed at that time was a basic oscillator that produces a waveform. But we didn't think of the instrument we were developing right away. The biggest decision that I think that I made had to do with... not interpretation—but dynamics, attack, decay, the articulation of music. He had wired the doorbell button up, so that it gave an attack to the note, he put a potentiometer on that, so that you could control the attack, and did the same thing with the decay. And so that was the beginning of probably the most important singular part of the original analog synthesizer.[6]

Many engineers and musicians feel the Moog synthesizer's best qualities come from its interior technology: electronically generated sounds driven by voltage-controlled transistor technology, organized into standardized modules, oscillators, filters, and a keyboard. In a series of circuit diagrams drawn by Bob Moog, the interconnections and switching points of modules that make up the core of the "Moog sound" have been flattened into silent schematic relationships on paper. Rendered in two dimensions, the circuit diagrams for the modules of pitch control, an envelope generator, a noise generator, and trigger extractor carry the visual weight of an abridged dérive. On the other hand, the complexity of the voltage-controlled oscillator module spreads out over several sheets of paper, meandering up and away from itself, seeming more like the template for the haphazard city where these psychogeographical walks might happen.

Once Deutsch returned home, Moog used reel-to-reel tapes as a correspondence method, noodling on the development and the sounds from the new device he called "the old Abominatron." He warned Deutsch: "It doesn't sound like much when I play it, but maybe somehow, someone with a bit more musicianship and imagination can get some good things out of it."[7] By October, Deutsch had the synthesizer in hand and began composing electronic music on what would later be elevated from "Abominatron" to "the first complete Moog prototype." The initial composition created on a Moog synthesizer was "Jazz Images: A Worksong and Blues." It is a decidedly apparitional composition, grafting 1960s free jazz onto traditional blues structures. Deutsch notes:

> I wanted to write a piece that was traditional in a sense but new jazz in another sense and new music in another sense, very often not tonal, very often sound-oriented, and definitely using the sounds of this new prototype instrument.[8]

From the first composition generated on its keys and modules, the Moog synthesizer was immediately infiltrated by portmanteaus of the past. If we are inescapably haunted by culture, and every*thing* is liminal and oscillating, the Moog makes an interesting case study. Even Deutsch, as early as 1981, had already called out the fallacy of the "newness" of synthesized sound:

> Apparently nothing stays new for long; or perhaps nothing is ever altogether new. Roger Bacon imagined the sounds of the Moog in 1654. In a remarkable forecast of Trumansburg and Buffalo developments and the implications they have for twentieth century music, he wrote in *The New Atlantis* about "...sound houses, where we practice and demonstrate all sounds ...harmonies which you have not heard ...quartertones and less slides ...diverse tremblings and warblings of sounds ...We have strange and artificial echoes ...and means to convey sound through trunks and pipes.[9]

And this notion of "differánce"—the disquieting nature of only being able to speak of hauntings *indirectly* through other things—has especially infiltrated the afterlife of the Moog. Mort Garson was one of the first electronic composer/arrangers to record a commercial album using the Moog modular synthesizer, collaborating with the musician Paul Beaver on *The Zodiac: Cosmic Sounds*, released in 1967. An instruction on the back of this record reads: "Must be played in the dark."

Garson's later efforts, *Black Mass Lucifer* (1971) and *Ataraxia: The Unexplained* (1975), attempted to capture the dimmest corners of occult history, meshing this once-taboo subject with the voltage-controlled sounds of the Moog. Even with song titles like "Voices Of The Dead" or "The Evil Eye," Garson's music today is less ominous, and perhaps more cheesy, considering its layers of proto-prog-rock noodling. Garson seemed to have his finger on the pulse of all things New Age, composing music for plants, music for the signs of the zodiac, and music for "sensuous lovers." His arrangements accompanied the most cosmic of all happenings in the 1960s—the background score for the CBS television transmissions of the Apollo 11 launch. "The only sounds that go along with space travel are electronic ones," Garson said. "[The music] has to echo the sound of the blastoff and even the static you hear on the astronauts' report from space. People are used to hearing things from outer space, not just seeing them."[10]

In the 1969 Kenneth Anger short film *Invocation of My Demon Brother*, a performed occult ritual—that is, in part, a satanic funeral for Anger's pet cat—appears as a psycho-sensorial wash of cryptic imagery. The harsh, repetitive soundtrack was provided by Mick Jagger. It consists of a grating patch that was produced on one of the first modular Moog synthesizers. Electronic sound became a mainstay in 1970s and 1980s horror films—a way to convey a sense of dread, seemingly inexpressible by any other musical means. Two decayed notes repeat like a death march in *Night of the Living Dead*'s nocturnal zombie feasting scene; the prog-rock band Goblin set the witchy tone for Dario Argento's giallo films *Deep Red* and *Suspiria* with a modular Moog; John Carpenter composed his own soundtrack for *Halloween* and *The Fog* with a Moog IIIP Modular system and a Minimoog.

Traces of horror aside, contemporary musicians, especially those who call the Ghost Box record label home, have continued to exploit modular analog synthesis. The artists working with Ghost Box share a common drive to find the "keynote" of sound that seemingly conjures post-industrial British landscapes, forgotten séance rooms, and parallel worlds. Opting to use a variety of analog synthesizers as a primary invocation tool, they retreat from compressed sound and the closed system of purely automated and digital music-making. Using patch cords, wires, and keys to shape sound is at once frenetic, random, and meditative. There is a sort of ritualistic power in commanding the signal, pivoting one tweak or another into becoming a presence, and then determining how it lives and expands or decays into noise. The Mount Vernon Arts Lab describe their LP *Séance at Hobs Lane* as "a lost classic of British electronics ... a psychogeographic investigation into a world of abandoned Underground stations, Quatermass, eighteenth century secret societies and the footsore reveries of a modern Flâneur."[11] And the Advisory Circle's *From Out Here* narrates the story of a computer becoming sentient, "where bucolic English scenery is being manipulated and maybe even artificially generated by bizarre multi-dimensional computer technology."[12]

In a 2014 oral history, Deutsch recounts the foundational moments in a life in music. A particularly powerful memory occurred in a strange form of aural synesthesia:

> I came from a very poor family. They ran a chicken farm on Long Island. There was a piano in the house, and it was a wreck. I can remember that. But I liked to just touch and play on it. I was three years old and I was out in the garage of the house. There was an empty unattached garage, not like modern houses.
>
> And I was standing there in the garage with a stick in my hand. And I remember distinctly that if I hit the stick on the ground, I heard a tone. And if I took that stick, and I moved it like that, up and down, I could hear (SINGING) "Dum, Dum, Dum, Dum, Dum." And I could start playing tunes on that stick. And then, all of a sudden, my brain came in, and it said, "That's not making any sound. It's just going, 'Thud, Thud, Thud.'" And it actually scared the devil out of me. I remember running into my

mother, in panic, thinking, "How did I do that? What am I hearing? I can't hear anything."[13]

My own connection to the Moog synthesizer (and its historical actors) is personal. The instrument was transferred from Deutsch's home studio to Hofstra University's electronic instrument studio, where he taught. While there, however, the instrument was underutilized: "[My] students would occasionally use this, but most of the time, they wouldn't. And when I talked to them about it ... they said, 'No. It's too important.'"[14] In *Keyboard News*, September 1982, Deutsch remarked on an auction where the Moog prototype went unsold, and his surprise at finding no interest at the Smithsonian Institute: "People should realize that someday it will be regarded somewhat like a Stradivarius," he said.[15]

Luckily, when Deutsch reached out to Robert Eliason, former curator of musical instruments at the Henry Ford Museum, Eliason understood the historical importance of the synthesizer, and he acquired it in 1983. Today, object number 1982.68.1 is partly under my purview as the curator of sound technology. And against all recommendations of playing favorites among children or museum object collections, it undeniably falls within that territory.

When Deutsch transferred ownership of the prototype Moog synthesizer to the museum, he gave a concert performance on the stage of the venue's historic Anderson Theater, effectively becoming the last person to play the instrument. In 2014, he returned, almost 35 years later, to give a follow-up performance, and to become the subject of an oral history. When Deutsch showed up, he handed me a DVD transfer of the original 1983 concert, videotaped, forgotten, and unseen by most. Purple letters produced by a character generator float over top of the opening scene: "From Whence the Moog?"[16] Deutsch is hunched over in the blurry video footage, playing the prototype Moog on the very stage where we would collapse the past into the present with our 2014 interview.

Asked how he felt about his reunion with the object, Deutsch said:

> When I walked into the theater today and saw the Moog prototype sitting on the stage, the first reaction I had was akin to seeing something somehow from your life. Now, I won't say it was like seeing a child, but it was the same kind of reaction—that it was a child. It was like a child in my memory, you know—in my musical memory.[17]

...

(BREAK IN TAPE)

Notes

1. Pinch, Trevor J. and Frank Trocco. *Analog Days: The Invention and Impact of the Moog Synthesizer* (Cambridge, Mass. Harvard Univ. Press, 2004), 23.
2. The conference was the New York State School Music Association.
3. Deutsch, Herbert. "On Innovation Interview with Herbert Deutsch," transcript of an oral history conducted 2014 by Kristen Gallerneaux, The Henry Ford, Dearborn, 14.
4. Deutsch 2014, 23.
5. Pinch, 22–23.
6. Deutsch 2014, 21.
7. Deutsch, Herbert A. "The Abominatron," *From Moog to Mac* (North Hampton, NH: Ravello Records, 2012). Sound recording.
8. Deutsch 2014, 28.
9. Deutsch, Herbert A. "The Moog's First Decade 1965–1975," *NAHO*, (Albany, N.Y: New York State Museum and Science Service: Fall 1981), 7.

10. McLellan, Dennis. "Composer was a synthesizer pioneer," *latimes.com*. Los Angeles Times, 11 Jan 2008 <www.articles.latimes.com/2008/jan/11/local/me-garson11> Accessed 6 April 2016.
11. "Mount Vernon Arts Lab — The Séance at Hobs Lane — Description." *GhostBox*. GhostBox, 2001 <www.ghostbox.greedbag.com/buy/sance-at-hobs-lane-0/> Accessed 6 April 2016.
12. "The Advisory Circle — From Out Here — Description." *GhostBox*. GhostBox, 2015 <www.ghostbox.greedbag.com/buy/from-out-here/> Accessed 6 April 2016.
13. Deutsch 2014, 3.
14. Deutsch 2014, 52.
15. "Moog Prototype Unsold at Auction," *Keyboard* (San Francisco, CA: Miller Freeman Publications, September 1982), 64.
16. *"From Whence the Moog?"* Dir. Unknown. Perfs. Herbert Deutsch. 15 January 1983. Unreleased video footage transferred to DVD, The Henry Ford Museum, 2014.
17. Deutsch 2014, 44.

DIRT SYNTH
Excavating The Sound Of The Poltergeist

The poltergeist, or "noisy spirit," is volatile, erratic, and interacts with purposeful malevolence. This is in opposition to the classic haunting, which is benign, repetitive, and often manifests due to a historical crises or familial connection. Hauntings leave no trace, but the "apportations" of the poltergeist are often elemental in nature, causing damage to dwellings: showers of stones or mud, indoor rain, spontaneous fire. In poltergeist cases, we see an unfailing narrative focus on the manipulation and *destruction* of material culture. Poltergeist activity has been hypothesized as being linked to humans. But, in the end, it's through the manipulation of material culture—and the latent power of objects within a given space—that all narratives evolve.

Above all, the poltergeist is perhaps best known for disembodied knocking on walls and furniture—the production of sound that witnesses habitually describe as coming from within the bones of dwellings and objects. In July of 1958, an Illinois family was "awakened by whistling noises from the refrigerator, crunching sounds from the wall and the shaking of their bed". Their 14-year-old daughter "was hit on the arm with a cabbage, and struck in the back of the head by a quarter-pound of butter."[1]

Whether legitimate or a hoax, poltergeist events consistently record interactions between the unseen, hyperlocal landscapes, mistreated objects, oppressive sonic effects, and the affected victims as a type of multilingual haunting. The overtly physical nature of the poltergeist usually prompts an investigation into the geophysical qualities of the site in order to rule out natural causes for the phenomenon. Seismic data, rogue electrical waves, and the presence of infrasound are checked against the uncertain energies of the poltergeist.

In 2006, I began gathering soil samples from the sites of purported poltergeist activity. Notable samples include dirt from the location of the Baldoon Mystery in Ontario, Canada, an infamous and very early case centered around a fire-raising spirit that ultimately led to the destruction of John McDonald's family home in the 1830s. A connecting sample was brought to me from Kirkcudbrightshire, Scotland. A fire-obsessed poltergeist suspiciously like the one at Baldoon was reported here and, in fact, was the same town where the Scottish emigrants who would later settle in Baldoon, Canada originated.

In the United States, I gathered samples from the Little Haiti district of Miami, where, in the 1960s, a poltergeist attached to a young Cuban immigrant took up smashing Florida-branded souvenirs in the Tropication Arts warehouse where he worked. Soil and pulverized gold medallion seed pods were gathered from the front yard of the innocuous-looking Craftsman house where the "San Pedro Poltergeist" terrorized Jackie Hernandez in 1989.

A decade after beginning to build this archive of samples, I began to experiment with methods of amplifying the sonic qualities of the individual soil samples. Initial experiments with contact microphones failed to produce much. Pushing the

---

[1] Loomis, C.G. "Illinois Poltergeist," *Western Folklore*. 17.1 (1958), 62.

electric ear around in tiny piles of dirt only produced homogenous crinkling clicks and pops. I found more success in the realm of analog synthesis. The individual modules that make up an analog synthesizer, which is more about twisting knobs and patching cords than pressing keys, are infinitely customizable, and allow one to strip back sound to its most basic qualities: waveforms, controlled voltage, resonance, attack and decay, and tone.

Using a combination of off-the-shelf and custom-built modules, I have built a synthesizer that is honed toward the amplification and processing of external sounds—and the smearing of those microsounds across longer time-signatures, which is known as granular synthesis. Patch cords lead out from the main instrument directly into the poltergeist dirt samples, in order to create self-generating drones. Since the synthesizer partially generates sound based on electromagnetic interactions with the content of the soil, there is a marked difference in the sonic qualities of different types of dirt that are influenced by the presence of iron deposits, limestone, sand, and moisture levels. When the open ribbon-circuit module is touched with the fingertips, the person playing the instrument literally becomes part of the circuitous electrical path for the sounding-off of the spirit soil. Finally, activating the time-warping of the granulator allows one to build the sounds into lush ambient landscapes, which echo back into themselves and create new feedback loops.

The intent here is to enter into a speculative collaboration between human and non-human elements, where the result is as unpredictable as the nature of the poltergeist, perhaps carried along in the generative microbial qualities of the dirt. If landscape can be charged or imprinted by the supposed supernatural events that once occurred at these sites, the dirt and the synthesizer are intended to act as an (admittedly, somewhat absurd) agent for sonic excavation and exorcism in the present day.

VI. ANCHORS

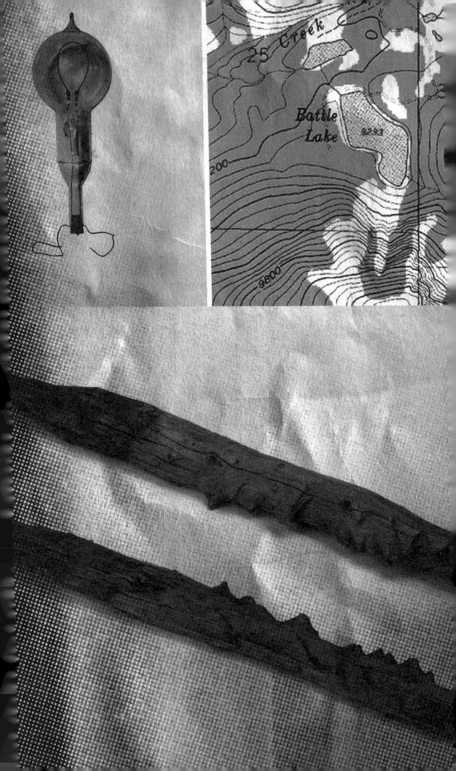

PROSPECTOR STAKE FROM A MYTHICAL LAKE

A wooden prospector's stake made of cedar, cataloged as a "pointed object." It's 23.5 inches long, but denies a consistent diameter measurement, with one side undulating in spiked peaks and valleys. The exposed, craggy top was blasted away—years peeled back—by sand-laden winds in Wyoming, blowing in hard from the west. Its natural waveform is dramatic. Over the course of 30 years, its eroded wavelengths emerged, busting outwards— visual tension moving away from its equally dramatic vertical cracks. The bottom third of the stake remains cylindrical, stained a darker shade of brown by the earth. Small pieces crumble off every time it's moved. If it was to be replanted in the ground, a strong hammer-blow to the top would probably reduce it to wooden shards, the smell of its locked-up cedar core reactivating in the violence of the swing.

    This stake was once planted in the ground at a fictional lake. Its wood was an eroded witness to the total solar eclipse of 1878. It exists in a web of inaccuracies and half-truths, yet the story of the marring of its surface is undeniably concrete. When the stake was donated to the Henry Ford Museum in 1931, it arrived with a story scrawled on a paper tag in cursive ink: "Fifty-two years ago Mr. Edison, in company with U.S. Army officers and Union Pacific officials, camped at the Lake, now known as Lake

Edison. While in camp there, Mr. Edison conceived the idea of the incandescent lamp. The Gold Rush occurred just ten years after Mr. Edison's visit and the stake I am sending you by Parcel Post was used at that time."[1]

Battle Lake is nestled into the crook of the Medicine Bow mountain range in Carbon County, Wyoming. The names of surrounding locations give a hint to the heartbeat of the region: Purgatory Gulch, Indian Bathtubs, Hell Creek, Bachelor Creek. The story of Edison's conception for the incandescent bulb at Battle Lake is apocryphal, yet the stroke of genius that lit up Edison's brain has become runaway mythology, often reported as truth.

On July 29, 1878, Edison *was* part of the "Draper Expedition," which gathered to witness a total solar eclipse, 75 miles north of Battle Lake, in Rawlins, Wyoming.[2] This viewing location promised to be predictably dry and clear, with optimal viewing for all 191 seconds of the eclipse. Edison's purpose among the Draper party was to test the function of his tasimeter, a device that measured subtle temperature shifts. Attached to a telescope, Edison believed it might reveal information about the mysterious energies of sun coronas, which were soon to be amplified by the eclipse.

When Edison arrived in Rawlins, he found the town overrun by eclipse-seekers, forcing him to convert a chicken coop into a temporary scientific observatory. Even as the totality of the eclipse was nearing, he struggled to align his delicate equipment against a strong windstorm. In the last few moments of the event, Edison's experiment was salvaged. The tasimeter swung wildly toward the red, indicating massive heat levels in the sun's corona.

A letter written by Robert M. Galbraith, dated November 10, 1929—over 50 years after the expedition—listed his fellow members for the "Eclipse Expedition." Galbraith, a mechanic for the Union Pacific Railroad, joined Dr. and Mrs. Henry Draper, who headed up the party, along with the English astronomer Norman Lockyer (co-discoverer of helium and founding editor of *Nature*); George Barker (dean of the University of Pennsylvania); Henry Morton (president of the Stevens Institute of Technology); a Professor Meyers of Iowa; several employees of Clark Brothers Telescope Manufacturers; astronomy professor Craig Watson (and Mrs. Watson) of the University of Michigan; and one unknown man.[3]

In this letter, it's revealed that Marshall Fox, a reporter at the *New York Herald*, was also present to witness celestial overlays. Galbraith perpetuates the myth of the "Incandescent Conception," claiming that Edison "described it at the Breakfast Table. Marshall Fox, on arriving at Rawlins [Wyoming], sent a 2000 word message to his paper on the subject. I heard later that he and the Associate Editor came near losing their jobs for publishing such rot."[4] A century later, the Wyoming board of tourism continues to hold fast to Edison's presence, building a monument in his honor and claiming, "While vacationing here in 1878, Edison threw a broken bamboo fishing pole in the fire and was intrigued by the way the frayed pieces glowed. These observations supposedly gave Edison the idea on how to develop his own design for the light bulb's filament."[5] In reality, the filament design was not plucked out of those campfire embers as a readymade idea—it took thousands of hours of labor, money, and experimentation at Menlo Park before bamboo was tested as a practical option.

After weeks of preparation and a few minutes of an eclipse lightshow, Edison and George Barker decided to travel south to Battle Lake for a hunting and fishing trip. The mosquitos and blackflies must have been thick enough to create their own high-frequency din in those woods. While camping at Battle Lake, near the spot where the prospector's stake was planted, Edison attempted to relax before returning to his hectic schedule of invention at Menlo Park. He was most likely dreaming about trapping his next meal, not capturing light in a zero-gravity glass bulb. Thanks to the uncertainties and fluidities of time and truth, the lake became a site that both glorifies and denies the reality of the "lone genius inventor." A case in point: Searching for a "Lake Edison" in Wyoming nets zero results. This is in contrast to the following claim from Galbraith: "I suggested that the name of the lake be changed to Edison Lake. They had me write Mr. Edison telling him of the proposition and just a few days ago received his reply that he would appreciate the honor."[6]

The name "Lake Edison," which is part of the tale scripted onto the tag connected to the prospector's stake, was never officially adopted. The honorific has been worn away by time, much like the cedar pole that was planted in the same spot just a few years after the eclipse. Cedar is an indigenous species, meaning it's likely that the stake was severed and harvested from a nearby tree stand. This is a case of living materials shifting taxonomies into something cannibalistic—an amputation now meant to signal not only legal boundaries, but the intent to harvest that site's own natural capital. When the stake was driven into the ground at the edge of the lake, it was done so based on claim speculation, attributed to the discovery of copper deposits in the area. Temporary settlements were established. Prospectors rushed in to reap the metals. The Rudefeha Mine was one of nearly 200 other working mines within travelling distance of Battle Lake, which experienced a brief heyday of industry.[7] The nearby, drolly named town of Encampment quickly peaked as a center for smelting copper and piddling amounts of silver and gold until everything went bust in the first years of the turn of the century.

Local folklore from the peak of Battle Lake's mining rush often recounts the story of a phantom miner who lost his life at Slaughterhouse Gulch.[8] This story is told nearly as often as the legend of Edison's "Incandescent Conception." The miner was the victim of a hastily placed dynamite charge and, for a time, he stalked travelers and spooked horses into upright heel-kicks on the trail that runs between Battle and Encampment. The maxim "there wasn't much left to bury" applied to the cause for the Slaughterhouse apparition, whose remains were dispersed throughout the Gulch. A story from the early 20th century might seem innocuous, but it apparently carried enough weight within its circle of witnesses to have endured. In the late 1910s, forestry workers enjoying dinner at a campfire saw a stone tape loop of the Slaughterhouse miner walking by. The forestry party was certain that the miner could be *nothing other* than a spectre, based on his failure to greet them at the edge of the fire, instead choosing to pass by with silent footsteps.

Something as simple as a wooden stake has the potential to haunt us, too, just like the half-true legends and ghosts born out of the violence of settlement. From its first encounter, this silent cedar stick has had an affecting presence upon me with an intensity that others find difficult to share. It has become the symbol of a "thinking-through thing," an entropic harbinger of a "thing not static," transubstantiated into a vessel for dialogue about nature and the fallacy of claims we sometimes lay over the natural environment. If objects contain "stored codes," their inherent "signals" could become scrambled.[9] So what if we were to begin this process of de-scrambling the true sense of place, time, and storytelling in a literal sense? By making the physical digital. By slicing and freezing (in an auditory sense) the most granular of samples generated by the visual call-and-response of cedar peaks abraded into shape by deep cellular time? What do those sound wave-like undulations exposed on the stake have to say, on playback? Sound is another way to recover the traces.

Notes

1. Letter from L.P. Pasewalk to Henry Ford. BFRC.REG.R229.File 31.884, Benson Ford Research Center, The Henry Ford, Dearborn, Michigan.
2. "Biographical Notice of George F. Barker," *The Popular Science Monthly* (New York: D. Appleton, September 1879), 695.
3. Correspondence from Robert M. Galbraith to George E. Howard, November 10, 1929. Edison Papers. Unprocessed. Benson Ford Research Center, The Henry Ford, Dearborn, Michigan.
4. Correspondence from Robert M. Galbraith to George E. Howard, November 10, 1929. Edison Papers. Unprocessed. Benson Ford Research Center, The Henry Ford, Dearborn, Michigan.
5. "Battle Pass Scenic Highway," *Travel Wyoming*. Wyoming Office of Tourism, 2016. <travelwyoming.com>. Accessed 3 August 2016.
6. Correspondence from Robert M. Galbraith to George E. Howard, November 10, 1929. Edison Papers. Unprocessed. Benson Ford Research Center, The Henry Ford, Dearborn, Michigan.
7. Larson, T.A. *History of Wyoming* (Lincoln: University of Nebraska Press, 1966), 334.
8. "Haunted Places in Wyoming," *The Shadowlands*, 1998. <www.theshadowlands.net>. Accessed 3 August 2016.
9. A tangential thought and model for an approach to artistic practice, inspired by Allan Kaprow's idea that "the *idea* of art cannot easily be gotten rid of (even if one wisely never utters the word). But it is possible to slyly shift the whole un-artistic operation away from where the arts customarily congregate, to become, for instance, an account executive, an ecologist, a stunt rider, a politician, a beach bum. In these different capacities, the several kinds of art discussed would operate indirectly as a stored code that, instead of programming a specific course of behavior, would facilitate an attitude of deliberate playfulness toward all professionalizing activities well beyond art. Signal scrambling, perhaps." See Kaprow, Allan. *Essays on the Blurring of Art and Life* (Berkeley: University of California Press, 1993), 104.

THE DOMES
Aural Fieldwork At The End Of The Line

In 1879, when the tracks to the Southern Pacific Railroad ran out, five overheated workers from Yuma dug their heels in the sand, and the community of Terminus, Arizona was born. The three buildings they erected in the aridity of the Sonoran Desert were truly the "end of the line." When the summer heat wave ended, the tracks from Terminus resumed to Tucson, and the micro-village grew into a small town. In a nod to the prehistory of the nearby Hohokam ruins—or perhaps to avoid christening the place as a future ghost town—railroad executives redubbed Terminus as "Casa Grande."

The town boomed as a railhead to mining prospects, but fires levelled the town center to cinders twice within a decade. When mining slowed in the 1890s, agriculture took over, and Casa Grande settled into itself, with a population of roughly 48,000. True to its railway roots, Casa Grande is alternately referred to as the "crossroads" and "heart" of Arizona. It is smack at the intersection of the I-10 and I-8 interstates, and almost dead-center between Phoenix and Tucson.

This type of backstory is typical. But at the southern edge of town, from an off-ramp on the I-8 near a Walmart distribution center, strange structures waver in the distance. Stuck in the scrabbly sand, on a small plot hugged by a landscape that is as sparse as Terminus must have seemed in the 1800s, a small UFO fleet and a space centipede appear to have crash landed on Earth. All of this alien strangeness is not an event that flew under the radar of the NSA. "The Domes," as they are known, are architectural ghosts—physical remainders of 1980s computing manufacturing gone wrong.

InnerConn Technology was incorporated in 1968 in Mountain View, California. The company was active at a feverish time in computing history, with hot competition on the mini- and personal-computing market. InnerCon, for their part, was one of many companies that produced circuit boards to fuel Silicon Valley's dreams. In 1982, the owner, Patricia Zebb, decided to transfer the company's headquarters to Arizona and build a new factory. This is the true origin of the Domes.

Over the course of six weeks, InnerConn commissioned four structures to be built using Thermoshell construction techniques. The buildings took the form of huge, balloon-like bladders, which were supported internally by steel trusses. Several inches of polyurethane foam were poured over the top, followed by several more inches of concrete. When the foam and concrete had hardened, the balloon and truss systems were removed, leaving behind an open space free of support columns—perfect for manufacturing assembly lines. The concrete and foam were passive insulation against the Arizona sun. There are signs of foundations laid for buildings that were never built, due in large part to InnerConn's techno-optimism being squashed when they went bankrupt in 1983. The company built an organic cathedral to computing assembly in the desert. Sadly, the company went bust before anyone had the chance to move in.

The Domes were doomed to never be repurposed. Exposed to the elements, orange polyurethane foam now pokes through the crumbling concrete. They are tempting to explore, though it's illegal to do so without permission. The acoustics created by their shape are a perfect match for their exterior otherworldliness. The site amplifies and feeds back upon itself—a kicked pebble sounds like a gunshot across a canyon. The reverb at the center of the largest UFO dome makes it sound like you are underwater.

There are rumors the Domes are used as rave caves and human trafficking drop-offs and lairs for satanic rituals. The graffitied interiors and smattering of beer cans on the ground make it more likely that "bored teenage hangout" is more accurate. But who knows what that eerie tapping and humming was that I heard coming through my friend's car air vents while we were parked roadside at the Domes. It might have been a ghost. Or the sound of a computer factory at the edge of the end of the world thrumming into action.

BLACK BRIDGE

K.'s memories of the day they visited the Black Bridge lack clarity, in parts, but it seems reasonable to allow such distant recollections to oscillate between reality and fiction.[1] A group trekked out there late one afternoon, near the end of summer. This part of the story K. knows to be true, because the memory is conjoined with a precise feeling of loathing associated with the last weekend of the summer, before returning to school.[2] She feels it even now, years removed from the cycles of an academic calendar. That day, K. and her friends were seeking one last gasp of adventure before the cycle began again—forced into winter basements filled with smoke and boredom and the full spectrum of teenage tension.[3] In retrospect, it's uncertain if you could call this a group of "friends." The gang of five or so was more like a collection of one- and two-off alliances. They weren't even from the same high

---

1 D1712.3. *Interpreter of dreams.* Jewish: Neuman.

2 D2145.2.1. *Summer magically lengthened.* (Cf. F162.1.1, F971.5.) Irish: Plummer lxxx, Cross.

3 F531.2.6. *Monster lies underground with trees growing all over his body. When his mouth is opened, man falls into it and is swallowed.* Finnish: Kalevala rune 17.

school; they were all thrown together by chance on that late August afternoon.[4]

K. was always the lone female of the group, so saying that "some of them dated" would mean, more accurately, "K. dated one of them" and later, "K. got bored, and dated another one of them."[5] Her anxious fervor to escape small-town life. K.'s head (described separately from her body, always) was a muddled and medicated mess back then—clinical depression treated by a subpar cocktail of 1990s prescriptions. Her still-forming synapses were unable to process these medications, somehow prescribed by a gruff Scottish family doctor without her parents' permission at the first mention of "feeling sad," at age 14.[6] Borderline poverty exasperated by rural stasis led to a few atom-bomb detonations of emotion. Submitting herself to the headlock of memory, a forced recall of that time, causes a tidal wave of cringe-inducing fragments. K.'s coming of age was awkward and messy. Aren't they all? Best to leave some things to rest.[7]

But one day, after an inadvertent encounter with a photograph of the Black Bridge, a visceral reaction occurred—an intense compulsion to dredge up old pasts, to give them legs, and to let them run rampant in the present.[8] Back then, walking across sunburnt grass, crisp under black one-star Chuck Taylors, and the sweep of a trench coat worn as an affectation

---

[4] J682.1. *Foxes desert their allies, the hares, when they foresee defeat by the eagle.* Wienert FFC LVI 48 (ET 64), 106 (ST 187); Halm Aesop No. 236.

[5] B163.1.1. *Wisdom from eating fox's heart.* Jewish: Neuman.

[6] H1064. *Task: coming laughing and crying at once.* (Rubs eyes with a twig to simulate crying.) DeVries FFC LXXIII 212.

[7] J21.5.2. *"Take side road rather than main one where three roads meet": counsel proved wise by experience.* India: Thompson-Balys.

[8] J2671.2. *Fool keeps repeating his instructions so as to remember them.* (He usually forgets them.) *Clouston Noodles 133.

in sticky humidity. Walkman in the pocket, play button set to cycle infinitely.[9]

The group trailed along a patchy dirt berm topped with steel towers, transmission lines plunging toward a fenced-in electrical substation. K. lingered behind, fingers trespassing through the wire fence, ozone up the nose, imagining the crackle of mutating voltage.[10] The group lumbered on, broad backs to her, passing through the trampled gateway of stinging nettles leading deep down into the brush. Even then, she didn't like to be doted on and didn't like for her gender to be thought of as a handicap, as an excuse for not being able to "hang with the guys."[11] K. had dropped all of her tenuous female acquaintanceships when she began high school. She removed herself from a decade's worth of catty hazing from elementary schoolgirls who couldn't relate to her (retrospectively funny) misanthropic descent into plaid shirts and resistance to makeup.[12]

That moment of fingers entwined in the fence, blasting telepathic SOS signals over the frayed lines, is stamped into place by the smell of sweet grass. A spontaneous carriage return of mellow relief pushed by on a gust of hot, dusty wind.[13] K. looked up to see the gang descending deeper into the screen of leaves beyond the trailhead. She jogged over to catch up, duck-footed, before anyone noticed her absence.

---

9 D789.6. *Disenchantment by repeating magic formula.* South Africa: Bourhill and Drake 237ff. No. 20.

10 F101.6.2. *Escape from lower world on horse of lightning.* India: Thompson-Balys.

11 D55.1.1. *Man magically stretches self to overcome cliff.* Hawaii: Dixon 91; Melanesia: ibid. 91 n. 104; Micronesia: ibid. 91 n. 105.

12 D11.2. *Woman transforms herself into a bird and lends her female organ to a boy. He fails to return it and she becomes a man when she resumes human form.* India: Thompson-Balys.

13 A1142.3. *Persons escape to sky and become thunder.* Calif. Indian: Gayton and Newman 96; S. Am. Indian (Chiriguano): Métraux RMLP XXXIII 177.

The brush gave way to a dirt trail, leading to the peppered gradient of grey railway gravel and greasy wood beams stained with oil from passing trains. The trail transformed into the Black Bridge. The group made their way toward the figure standing at its end. Like the keeper of Dante's own particular Hell, there is always someone standing on the graffiti-ruined trestles of the Black Bridge.[14] Even now, someone is probably there—a residual or real haunting, waiting to give a lecherous side-glance to any and all females, warning of the Criminally Minded Corporation gang who hold parties in the woods, or the bloodstained murder spot where "that kid" was killed (pointing to the wrong place, never saying the victim's name out loud), or offering to sell you a dime bag, which might turn out to be dried jimson weed picked from the side of the tracks. Only now it might be meth. No matter the time, you still have to pay the ferryman to enter the forest.[15]

When K. was 15, a boy from her high school lured a younger, seven-year-old boy named Daniel Taylor into an abandoned building just beyond the Black Bridge. How that kid came to be alone in the middle of nowhere is a mystery. Whether he was wandering for adventure as children in small towns do—as they all did—or whether he was chased into the forest, it didn't end

---

14 F91. *Door (gate) entrance to lower world.* Irish: O'Suilleabhain 33, 58, Beal XXI 311, 323; Gaster Oldest Stories 152; Jewish: *Neuman; New Zealand: Dixon 73; Africa: Werner African 184.

15 F92.2.1. *Girl gathering flowers swallowed up by earth and taken to lower world.* Greek: Fox 227 (Proserpine); Oceanic (Mangaia [Cook Group]): Dixon 74.

well.¹⁶ When Taylor's body was discovered a day after his parents reported him missing, the incipient sinister nature of the county was finally set loose. All that was vaguely gentle about its façade finished crumbling. In a place where murder on the whole was rare, this death ripped through the community. Its circumstances shocking and violent enough to call for lobbying reform in Parliament to amend the juvenile criminal code. Lovecraft's adage that the quaintness of isolated towns is just a scrim screen to hide lurking horrors holds true; Kent County's shadow had become just as long as Lovecraft's fictional Innsmouth.

This talk of the spectre of Taylor, whose final architectural embrace laid somewhere beyond the train tracks, was driving everyone deeper into the bush that day. One year after the events, they were legend-tripping toward idiocy.¹⁷ It was as if, by seeing the crime scene, they would reconcile their own feelings about their town. That its unsettling energy would somehow account for their own lack of understanding of how something so vile could be committed by one of their peers—someone who went to their high school, someone who shared the same routes through the halls, shuffled toward the same classes.

K. broke free from the one-point perspective of the train tracks and off to the left, teetering at the edge of what was obviously private property. It might have been a station house and storage yard for the Canadian National Railway. But there was no security, just an empty chain where a dog should be and a house with rotting lace curtains. There was a hush among the group—if anyone was actually in that house, they would go no

---

16 J21.36. *"Do not go alone on journey": counsel proved wise by experience. Helpful crab saves from attack by crocodile.* India: *Thompson-Balys.

17 D2121.2. *Magic journey with closed eyes. Person must not open eyes while on the journey.* (Cf. †C300.) *Chauvin VII 59 No. 77. 102 No. 376, VIII 148 No. 146; Hartland Science 174 (Japanese).—Irish myth: Cross [...] Boas RBAE VI 629.

further.[18] Calls would be made to parents. And the openness of the property made it impossible to shield themselves from view. A flurry of silent hand-signals suggesting abandonment soon had them moving down the dirt driveway and merging with the tree line, safe again, their talk muffled by the growing density of what had, by that point, transformed into true Carolinian forest.[19]

Liminality, *communitas*, and all of that Turnerian folklore-speak about rites of passage are not mere literary tricks; they actually do occur in moments of transformation, rooted entirely in reality. Bodily rituals. Moving in and through a space made taboo by the touch of horror and power.[20] Obstacles to overcome fear—courage, community, submitting to loss, and *being* lost—are archetypes that crept off the pages of ritual studies texts and into the real world that day.

A few of the guys lingered at some distance from the main group, already spooked by the emerging view of a crumbling cinderblock building with a swayback roof.[21] Reluctant to go near it, they voiced that there was no real reason to see it again. K. circled around to the front, past rusting oil barrels with tumorous leakage turned solid, past rank Queen Anne's lace, and shattered brown glass from chubby beer bottles. A section of blocks, about the height of an average adult, was knocked out of the wall, like teeth smashed out of a mouth, forming a jagged maw through

---

18 N685. *Fool passes as wise man by remaining silent.* Pauli (ed. Bolte) No. 32; Spanish Exempla: Keller.

19 D457.4.2. *Transformation: hair to forest.* *Fb "hår" I 771b; Jewish: Neuman.

20 F979.2. *Leaves of tree open and close to give the dead passage.* Irish: Plummer cliii, Cross.

21 E593.3. *If no lamp is lighted in a house for a period of fourteen days, ghosts take it for their dwelling.* India: Thompson-Balys.

which to pass. Although the wall was essentially transparent, no one crossed the threshold. The contents of the building were non-existent, except for one greasy vinyl office chair, dead center in the space, a beam of sunlight shining down beside it through a hole in the roof.

A length of faded yellow tape stuck on a window frame was blowing inward, activating the space in an uncomfortable way.[22] K. refused to acknowledge that it was most likely a leftover from the crime scene. Rising and twisting with each burst of wind, it manifested the imagined movement of a shifty spirit fleeing toward the corner, locked on repeat. Whatever the building remembered—whatever had been retained of the horrible acts inside it—would continue to play themselves out until the tape rotted away.[23] Ignoring the horror of her own superstitions, K. reached inside the window frame and loosened the tape from its catch. The synthetic anchor drifted down, deflated and defeated. It was her minuscule act of mercy and mourning, witnessed by no one.[24]

However, the metal office chair remained, and it was bone-chilling. The chair had already established a strong sense of runaway legend.[25] Townies proclaimed it to be the site where the murder happened. Daniel Taylor had been tied to a chair in the last moments of his life. There was no other detritus to indicate

---

22 E436.3. *Bats flying in house sign of ghosts.* North Carolina: Brown Collection I 677.

23 D1318.5.4. *Speaking blood reveals murder.* Frazer Old Testament I 101; Fb "blod" IV 47ab; Hdwb. d. Märchens s.v. "Blut".

24 D1318.5.3. *Each drop of innocent blood turns to burning candle.* English: Child I 172, II 39b.

25 E539.4. *Ghostly chair in cellar jumps up and down on three legs, points with fourth at spot on floor. Witnesses dig up body from under floor.* U.S.: Baughman.

anything had ever happened there.²⁶ The place was full of ghosts, but all the same, there were none to be seen. The perpetrator of the crime had a head full of demons known only to him, which he chose to exorcise in the most-vile way.

Apparently, the 17-year-old murderer, James Morin, had a steel plate in his head, the result of a vicious pummeling he received from his father. It obviously wasn't the first. Uttered like a mantra, after the fact: "He wasn't... right." Rumors that Morin had acted out scenes from the 1988 horror film *Child's Play* during the murder seemed logical. K. stifled something in her throat, then and even now—the collapse of rumor, history, violence, and place. She swallowed it and moved deeper into the bush, away from the pull of the building's vortex.²⁷

―――

There were concrete pillars painted with the usual "Smoke/Toke," "420," and metalhead allegiances to "Anthrax," "Maiden," et al., peppered throughout by the less penetrable cryptography of gang graffiti. The clear aesthetic differences between rebellious teen graffiti and the hurried territorial scrawl of gang-writing indicated the group was deep into the playground of the CMC gang. This was the gang everyone talked about, but never saw. Stories in the local newspaper warned residents to be wary of teenagers wearing combat boots and to glance with caution at their color-coded laces. So, K.'s preference for high Doc Martens, bottle-black hair, and pathological silence resulted in her being given a wide berth by other students as she walked the halls, head down, making it unnecessary for her to interact with anyone—a goal achieved.

―――

26 D1318.5.4. *Speaking blood reveals murder.* Frazer Old Testament I 101; Fb "blod" IV 47ab; Hdwb. d. Märchens s.v. "Blut".

27 E489.9. *In land of dead the dead walk on grass without bending it and on mud without sinking.* (Cf. †F973.2.) Chinese: Graham.

Black Bridge 253

Few members of K.'s group owned computers or had email addresses. No one had a cell phone. Yet they were all digitally aware. Upon finding one another, these cyberpunk music geeks and socially anxious weirdos abandoned their computer screens in favor of IRL interactions—at least for that summer. Together they whiled away an afternoon, backs to the ground, eyes to the clouds, laying on the top of the hill in Paxton Park.[28] That place was once a trash heap, sowed over with grass. A midden that might have easily been misinterpreted as a native burial mound.[29]

The mysteries of the forest kept unfolding that afternoon, each layer uniquely sinister, but never as corrupted as that murderous outermost passage. A brick smokestack tapered down into a lopsided brick building, one wall lined with ornate cast iron ovens for making more bricks. The trill of killdeer in the trees, cicadas buzzing their late-summer death appeal. Abandoned tunnels they vowed to revisit with flashlights. It's probably good that they never followed through with that adventure.

They emerged into a cleared patch that was the graveyard for a rusted-out truck and some kind of heavy operating machinery, the specifics now lost to time. On the back of a flatbed trailer, level with the tops of goldenrod, there were once-orderly piles of disintegrating paper rectangles. Not understanding what they were looking at, the oldest in the group explained that they were computer punch cards. Their markings—"Do not fold, spindle or mutilate"—seemed somehow cruel, given their location. The idea that thoughts were processed in and through those holes by people their parents' age, the same generation that would

---

28 F61.3.1. *Ascent to upper world in smoke.* India: Thompson-Balys; Caroline Islands: H. Damm Zentralcarolinen (Hamburg, 1938) II 88; New Hebrides: C. B. Humphreys The Southern New Hebrides (Cambridge (Eng.), 1926) 98; Yap: W. Müller Yap (Hamburg, 1918) II 666, 685, 695.

29 F402.6.2. *Demons live in waste mound.* India: Thompson-Balys.

have raised and failed to protect the innocence of Taylor. And Morin too.

Season by season, the iconoclastic Black Bridge brush was turning concrete intangibles back into sawdust. Those heaps of computer cards in the forest were reverting back into feed, flecks of black mold in stacked layers. Earwigs were just waiting to explode with the disrupting poke of a finger.[30]

And then there was "The Barge," rising out of the Thames Riverbank. It towered above the group at an unnatural diagonal, so high they didn't dare try to climb aboard. It was a rusted-out hulk run aground in the early 20th century, left to rot.[31] After it was abandoned, it became the site of another supposed tragedy—a teenage suicide. Little was known about the boy, other than that he used to gather with friends to play Dungeons & Dragons on the ship's deck.[32] It happened in the mid-1980s, but some knew a similar story from the next town over: A boy their age, the son of K.'s optometrist, was also obsessed with heavy metal and D&D. He was messing with his father's gun and shot himself. Some claimed the accident was purposeful, a way to rebel against the strictures of his father's ex-military discipline. In later years, a friend who made experimental short films on a heavy portable VHS recorder showed K. footage of that same boy giving a tour of his father's gun collection. Static lines rolled up the screen to reveal him aping for the camera, gun

---

30 D1413.20. *Magic earth-mould holds person fast.* (Cf. †D935.) Irish myth: Cross.

31 F931.5. *Extraordinary shipwreck in calm weather.* Icelandic: *Boberg.

32 D454.10. *Transformation: ship to other object. (Coffin)*

to his head, wild look in his eyes, bottle of beer in hand.³³ What did they think they were doing out there? Why exorcise these memories onto paper?³⁴

After paying homage to the Barge, they wandered deep into the forest, sitting on a fallen maple branch to rest and smoke stolen cigarettes.³⁵ The day transformed into explosions of orange, then the cold blues of dusk. As dusk turned into full-blown night, someone suddenly wondered whether they'd be able to find their way back home. K. vividly remembers what followed—stumbling through the forest at night, lighting sheets of a newspaper on fire to pierce the densities of the forest with fleeting light-bursts.³⁶ Flame spells to banish the bad energy of the day. Waltzing in panicked circles. Feeling the prickle of spirit fingers on their shoulders. Haunted by the place, and themselves.

Purple night, black canopy, wire vines, silver branches. Feeling totally fucked, like they were being absorbed into the woods and would never escape, their voices were muffled by the encroaching darkness, becoming lost to one another.³⁷ And then, a sudden breakthrough. A clearing. They walked silently back to the berm, back to the power station. At the cemetery, they sat to rest their quivering knees, realizing—then and now—that they had somehow walked past the bridge without ever having crossed it.

---

33 E389.2. *Summoned ghost audible and visible only to person who has summoned him.* Jewish: Neuman. Cf. Shakespeare Hamlet ("ghost scene").

34 E538. *Ghoulish ghost objects.*

35 D469.2. *Transformation: smoke to bridge.* Africa (Vai): Ellis 191 No. 8.

36 D2158.1.4. *Magician opens his eyes and forest burns for twenty-four miles in front of him.* India: Thompson-Balys.

37 D1368.5. *Magic forest seems to stretch farther as mortals travel within.* (Cf. †D941.) India: Thompson-Balys.

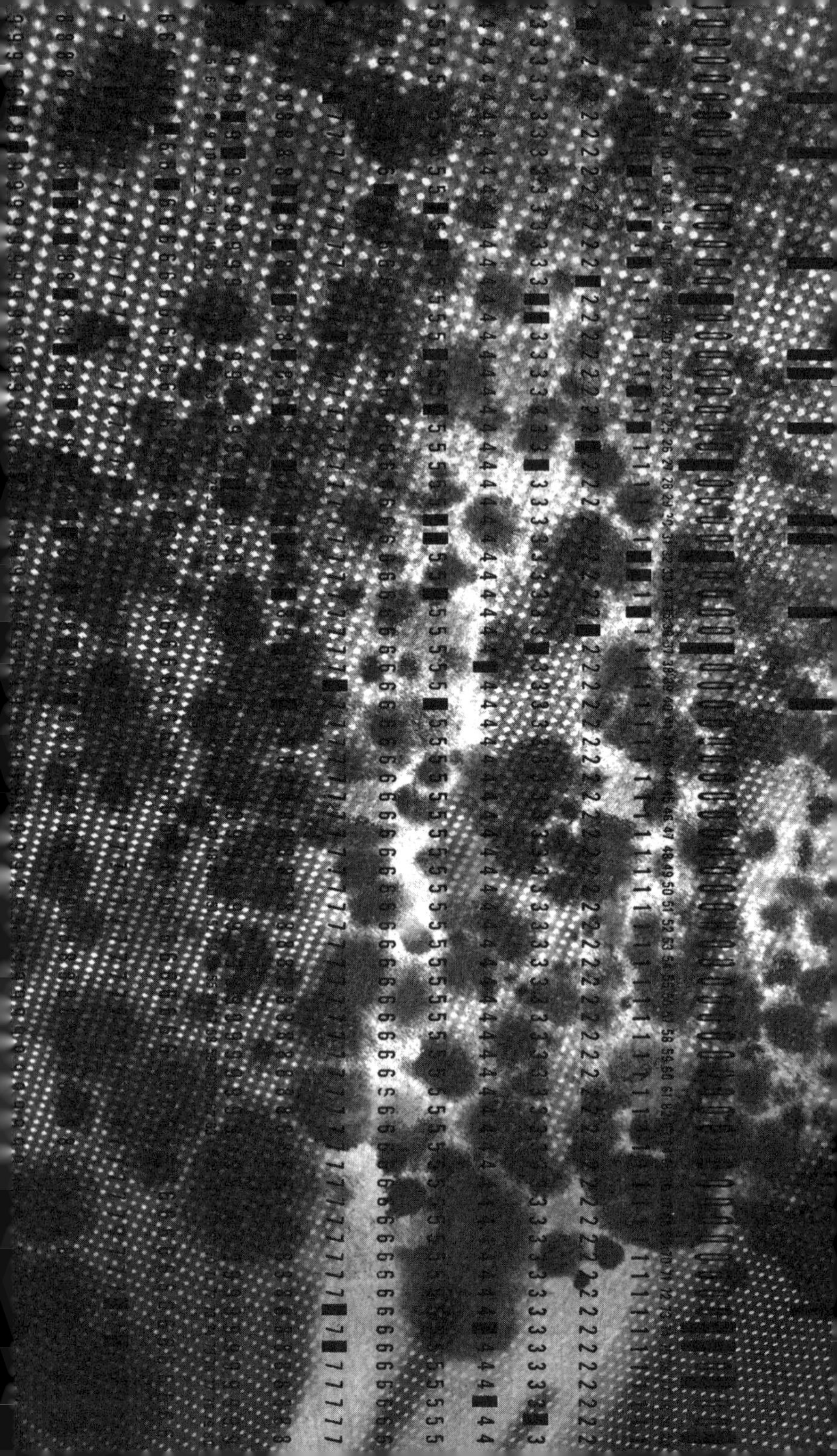

Throwing the Apparition of a Brick Through a Real Ghost: A Conversation with the Author

In which Kristen Gallerneaux discusses rare toasters, "imprints of defeat," witch germs, and growing up in a family of spiritualists in rural Canada

by Dave Tompkins

Can you talk about treating your hearing disorder with plankton?

I have this weird hearing condition called Ménière's, which makes the world sound like it goes underwater a few times a day. I've become pretty adept at externally masking when the internal bathysphere descends, even though it's pretty disorienting. It's why I don't drive. One time, the doctor filled my ears with algae casting material to make custom earplugs.

Your book title is a reminder that the deceased often have dodgy phone service.

One of my husband Bernie's relatives once received a phone call from her deceased husband on her cell. Seamus McShane, ringing in. But there was no static, only dead lines. For a few months after he passed—about five years ago now—everyone kept seeing him in crowds. I saw him at the mall. Bernie's sister saw him at the airport.

Bernie also has a terrifying childhood memory of being awakened by a bat trapped in the window screen next to his bed. How does your own childhood inform your research?

I grew up in rural Canada. My brother's best friend's dad, Lenny, was an archaeologist who was missing several fingers. How he lost them depended on the day: defusing a bomb in WWII, a snapping turtle, a gambling debt. He claimed to have worked for "CIA agents" on "UFO crashes" over Lake Erie in the 1980s and had suspicious "experimental metals" (likely tinfoil) in his shed that he "liberated" from the crash site. One of his hobbies was collecting chicken bones so he could build elaborate sailing ships with them. Another was embarrassing shy girls with his boxful of raccoon baculum. I think Lenny might be one of the reasons I seek out the humor and imaginative qualities lurking in the sub-basement levels of history.

Did you see any of the wishbone ships?

They were incredibly detailed, white-bone ghost ships, with calcified sails and all. They used to be lined up on the fireplace mantel—a fleet of phantom chicken vessels. If you were to stumble onto their farm, unaware of how nice the family was, you might think you'd run into Leatherface, with all the animal hides, pickled beings, and bones laying around.

You grew up in a family with three generations of spiritualists, eavesdropping on séances.

I remember once, at a family dinner, my uncle ran across the street to his apartment to recover some famous ectoplasm images from the early days. Foggy breath suspended in front of a Mary in a Bathtub statue in the front yard. But there are other images that are less easily explained. The supernatural stories are hinted at in the interludes throughout the book, cobbled together from a mess of semi-autobiographical memories. Ghost stories permeated my everyday life while growing up. But actually, it can become annoying as an adult to have that filter applied over everything at family gatherings. Sometimes, a dream is just a dream, and not an ill omen. I try to keep a healthy skepticism about things, in the agnostic sense of the word—and not an atheistic sense. Something one of my colleagues said once hit the nail square on the head: "I don't believe in ghosts, but I've met them plenty."

Was your helicopter-mechanic aunt more of a realist?

My aunt was a helicopter mechanic in Canada in the 1970s. She spent most of her career working up near decommissioned, deep Arctic DEW line sites, in a place called Tuktoyaktuk, in the Northwest Territories. She always said it was too cold for cold germs to survive up there. The tundra map looks like Swiss cheese. She relayed rumors of subterranean UFO bases up North well before I saw my first episode of *Unsolved Mysteries*. She also worked in a place called Qausuittuq, which translates to "Place with No Dawn." Incidentally, she lived with uninterrupted daylight and darkness, and had to use blackout curtains so she could sleep. From Google Earth, all of those places look like homes for The Thing.

You have some Native American Huron lineage, which also runs deep in Michigan.

Yes, I am Métis and have a strong Huron and Ojibwa lineage, and ancestors with names like Catherine Annennontak and Chief White Crane. This part of my ancestry was hidden for several generations, which is troubling and frustrating in all kinds of ways.

You once said, "Manitou zapping computer slugs are the only two minutes of cinema that matter in the world." Could you elaborate?

I'm really intrigued with this film *The Manitou* (1978), about an irate Algonquin entity named Misquamacus. That movie has floating grannies, séances, lasers, a cosmic inter-space dreamscape battle that involves finger lightning blasts and a 1970s-mainframe computer, plus skinwalkers!

Has your work as a trained folklorist given you a different perspective on your spiritualist upbringing?

I've always been interested in studio practice as an artist as much as the research and writing side of things. When I finished my art degree, I went into the folklore program at the University of Oregon. It allowed me to have the time and creative range to write about the types of things that academia usually frowned upon—paranormal culture and the like. My folklore mentor, Dan Wojcik, is an expert in eschatology and outsider art, and he let me have free rein. I also had a gig working in the Mills Folklore Archives, where I helped people access everything from the Bigfoot files to folk dance history to … menstruation lore. Yes, that's a thing.

How did you become the world's leading expert on rare toasters? You may be the sole owner of a Buck Rogers toaster.

My first gig working at a museum was to research a collection of over 600 electrical toasters, from the first commercially successful model—made by American Beauty and invented in Detroit, no less—to a modern-day Michael Graves design for Target. There isn't really a definitive book on toaster history, so I spent a lot of time looking at patents and pulling together the lost pieces of toast history. I was really excited to find a double of my favorite toaster in an antique store, an Art Deco General Electric model with a spider web embossed on the front. I bought it, obviously.

You've been pretty into salt too.

Here's a good one I came across recently in a book from 1898 titled *The Magic of the Horseshoe*: "In some regions a new tenant will not move into a furnished house until all objects therein have been thoroughly salted, with a view to the destruction of witch-germs." WITCH GERMS, man.

"A view to the destruction of witch germs": I want that postcard. It belongs in a national park registry. Actually, I think the world needs more witch germs. You're in charge of a lot of dead media artifacts at the Henry Ford Museum. What have you found lately?

I keep hoping to turn up a vocoder here, but no luck. We have this equipment that was from the old Telefunken wireless station in Tuckerton, New Jersey. It seems to be some kind of photo-sensitive paper (coated in a mild cyanide solution) in a light-tight box. I think it was used to record sound waves from radio receivers, but it looks like a turntable of chemical death. I also found some parabolic microphone dishes—some of the first. Looking very much like they lost their way from *Cat-Women of the Moon*.

What about the strobe gun?

Some of my favorite items are the most enigmatic. We have an item cataloged as a "photo radio gun" that was built by John Hays Hammond Jr, the "father of radio control." He made early remote-guided torpedoes and an autonomous "ghost ship" that sailed around the harbor, unmanned, in Gloucester, Massachusetts. Hammond was an eccentric. He had séances in his basement—complete with a Faraday cage—and also made radio-controlled pet dogs that you could tempt with fluorescent light-tube weenie wands. In real life, he preferred cats.

You and Bernie, a native Detroiter, often explore the city.

I've had super-vivid dreams lately of riding a bike through areas of Detroit that don't exist, and sometimes finding cellars full of magic botanica candles, sometimes old reels of film from dreamland monster movies that I wish existed. There was some really intense graffiti in Detroit a few years back; someone wrote "It Don't Exist" over and over on about a block of abandoned buildings. And then someone else blocked out the scrawls with rainbow-palette discount paint to make the nothing more (or less) invisible.

I like the plan to sneak into the abandoned Pontiac Silverdome, in hopes of picking up some psychic residue from Lions games. (Speaking of "imprints of defeat," it's too late—the stadium has been demolished.) Or maybe catch a spirit bleed from when the Fat Boys opened up for Wham! in 1985. Incidentally, the bookmark for my copy of **Mangled Hands** *is the obituary of Detroit Lions running back Mel Farr, who sang backup vocals for Marvin Gaye on "What's Going On."*

A few months ago, I found three photos of a man in a Lions shirt juggling bananas with a cardboard box over his head. They were stuffed in an old sketchbook. I have zero memory of where these images came from. I can imagine a Silverdome recon and a night drive with filtered red flashlights and the 80-pound reel-to-reel recorder that was used in family séances. Too bad the Dome is gone now.

Elaborate on this line you wrote years back: "Teeth-gnashing Inuit spirits that play football with children's skulls in the sky."

You're talking about aurora folklore. That line relates to the infrasound phenomena called "The Hum." [For more, see the chapter "Humming Through Salt: AUDINT's Spurious Frequencies"] Anyway, people sometimes experience sound effects when they see the aurora borealis lights. Crackling radio static and whispers. Scientifically, the sound can be explained away as effects of infrasound and VLF waves. But from a folkloric perspective, this phenomenon has inspired centuries of northern folklore. I tried to do an art-science collaborative project with the nanoengineering department at my last university, to try and counter the effects of the infrasound from the Windsor Hum. I wanted to make a nanotech impregnated spray paint for people or houses or whatever that could counter the physical effects of the external infrabass bleeding into people's lives. Apparently, there is a lot to learn from 19Hz and below. But the scientists were not into it.

One day at work, you photographed a joystick encrusted with radioactive grime from New Mexico. Y'all wore masks that day.

I talked on the phone with the guy who runs the landfill in New Mexico where "The Atari Burial of 1983" happened/ didn't happen. The legend is true. The Atari Dig was a modern archaeological excavation, perhaps the first-ever e-waste dig, to uncover the legendary "Atari Tomb." At the time, the company was going bankrupt, so they emptied one of their warehouses full of unsold game merchandise—roughly 700,000 titles, including the infamous *E.T.* game from 1982—and buried them in New Mexico. A bunch of Atari and pop-culture enthusiasts banded together, along with people who were present on the day of the 1983 burial, to try and find the tomb. The dig team succeeded, despite being blasted by a major desert dust storm.

When I spoke to the owner of the landfill, he told me, "You know you're getting a REAL ONE when you open the box and it smells like it was buried in a dump for 30 years." And then he started yelling—because, you know, he works at a dump, and there are machines around. "THEY SMELL REAL BAD, BUT THEY'RE SAFE, DON'T WORRY... I HAD THEM TESTED!" He said he and his friends stole a bag full of cartridges in 1983 instead of burying them. They drove around New Mexico with their car windows down and tossed them on everyone's lawns like newspapers.

So, these excavated E.T.*s will end up in the Henry Ford archives, keeping company with the monkey bar. [See Introduction for more on the monkey bar.]*

When I showed someone the monkey bar a few days ago, they said it seemed like if you stared at it too long, you'd end up getting sucked in, like the photo at the end of *The Shining*, and trapped inside one of the monkey bodies, in monkey hell.

It's like that Shirley Jackson story about the girl trapped in the painting. As the last one out of the office, do you ever stop by the monkey bar on the way?

Here are some instructions I once received for the end of the workday: If you're the last one here, flick all the light breakers off by that terrifying horse statue. But make sure you have a flashlight to find your way back. Maybe the monkeys are writing their own story whenever the lights are off in storage. There's a tiny manuscript waiting to be edited, stowed under the lid of the piano.

BIBLIOGRAPHY

Apgar, Charles E. "How I Cornered Sayville," *The Wireless World* 3, no. 32, November 1915.

AUDINT. *Dead Record Office*, New York: AUDINT Records 2014.

Barrett, William F. *Psychical Research*, London, Williams & Norgate: 1911.

"Battle Pass Scenic Highway," *Travel Wyoming*, Wyoming Office of Tourism, 2016. <travelwyoming.com>. Accessed 3 August 2016.

Beauchamp, Ken. *History of Telegraphy*, London: The Institution of Electrical Engineers, 2001.

"Biographical Notice of George F. Barker," *The Popular Science Monthly*, New York: D. Appleton, September 1879.

Bjorkman, Edwin. "The Music of Francis Grierson," *Harper's Weekly*, LVIII, February 14, 1914.

Blundell, Stephen. *Magnetism: A Very Short Introduction*, Oxford: Oxford University Press, 2012.

Bpoag. "I believe I know who was behind the 'Max Headroom Incident' that occurred on Chicago TV in 1987." *Reddit*, Reddit, Inc. 1 December 2010. <www.reddit.com/r/IAmA/comments/eeb6e/i_believe_i_know_who_was_behind_the_max_headroom/> Accessed 30 March 2016.

Bpoag. "New developments in the Max Headroom Incident mystery!" *Reddit,* Reddit, Inc. 11 October 2015. <www.reddit.com/r/IAmA/comments/eeb6e/i_believe_i_know_who_was_behind_the_max_headroom/> Accessed 30 March 2016.

Braga, Newton C. *Electronic Projects from the Next Dimension: Paranormal Experiments for Hobbyists,* Boston: Newnes, 2001.

Britten, Emma H. *Nineteenth Century Miracles; or Spirits and Their Work in Every Country of the Earth: A Complete Historical Compendium of the Great Movement Known As "Modern Spiritualism,"* Manchester: W. Britten, 1883.

Britten, Emma H., and Margaret Wilkinson. *Autobiography of Emma Hardinge Britten,* Manchester: John Heywood, 1900.

Bromberg, Joan L. *The Laser in America, 1950-1970,* Cambridge, Mass: MIT Press, 1991.

Brooks, John A. *Britain's Haunted Heritage,* Norwich: Jarrold, 1990.

Brown, Patrick R.J. "The Influence of Amateur Radio on the Development of the Commercial Market for Quartz Piezoelectric Resonators in the United States," *Proceedings of the IEEE Frequency Control Symposium,* 1996.

Brown, Rosemary. "Rosemary Brown in Conversation, 1973," *Okkulte Stimmen—Mediale Musik: Recordings of Unseen Intelligences 1905-2007,* Berlin: Supposé, 2007.

Clark, Mark and Henry Nielsen. "Crossed Wires and Missing Connections: Valdemar Poulsen, The American Telegraphone Company, and the Failure to Commercialize Magnetic

Recording," *The Business History Review*, The President and Fellows of Harvard College, Vol. 69, no. 1, Spring 1995.

Colvin, Barrie. "The Acoustic Properties of Unexplained Rapping Sounds," *Journal of the Society for Psychical Research*, 73.2, no. 899, 2010.

Connor, Steven. *Dumbstruck: A Cultural History of Ventriloquism*, Oxford: Oxford University Press, 2000.

Connor, Steven. *Paraphernalia: The Curious Lives of Magical Things*, London: Profile Books, 2011.

Connor, Steven. "Ears have Walls: on Hearing Art," *Sound*, ed. Caleb Kelly, London: Whitechapel Gallery, 2011.

Crane, Clare. "Jesse Shepard and the Villa Montezuma," *The Journal of San Diego History* 16.3, 1970. <www.sandiegohistory.org/journal/70summer/shepard.htm> Accessed 30 January 2016.

Crane, Clare. "The Villa Montezuma as a Product of its Time," *The Journal of San Diego History* 33.2-3, 1987. <www.sandiegohistory.org/journal/87spring/villa.htm> Accessed 30 January 2016.

Crouch, Tom D. *The Eagle Aloft: Two Centuries of the Balloon in America*, Washington, DC: Smithsonian Institution Press, 1983.

Crowley, David, and Jane Pavitt. *Cold War Modern: Design 1945–1970*, London: V & A, 2008.

Crystals go to War, Reeves Sound Lab, 1943. Prelinger Archives. <www.archive.org/details/prelinger>. Accessed 28 January 2016.

Cubitt, Sean. *The Practice of Light: A Genealogy of Visual Technologies from Prints to Pixels,* Cambridge, Mass: MIT Press, 2014.

Daukantas, Patricia. "A Short History of Laser Shows," *OPN Optics & Photonics News,* The Optical Society, May 2010.

Davis, Erik and Michael Rauner. *The Visionary State: A Journey Through California's Spiritual Landscape,* San Francisco: Chronicle Books, 2006.

Davis, Erik. *Techgnosis: Myth, Magic + Mysticism in the Age of Information,* New York: Three River Press, 1998.

"Delphic Panaceas: An Infrasound / Ultrasound and Video Installation," *Artcite.* <www.artcite.ca/history/2015.html> Accessed 28 January 2016.

Derrida, Jacques. *Specters of Marx: The State of the Debt, the Work of Mourning, and the New International,* New York: Routledge, 1994.

Deutsch, Herbert A. "The Moog's First Decade 1965-1975," *NAHO,* Albany, NY: New York State Museum and Science Service: Fall 1981.

Deutsch, Herbert A. "The Abominatron," *From Moog to Mac,* North Hampton, NH: Ravello Records, 2012.

Deutsch, Herbert. "On Innovation Interview with Herbert Deutsch," transcript of an oral history conducted 2014 by Kristen Gallerneaux, The Henry Ford, Dearborn, Michigan.

"Digital Television," Federal Communications Commission. Updated 7 December 2015. <www.fcc.gov/general/digital-television> Accessed 30 March 2016.

Dyson, Frances. *Sounding New Media: Immersion and Embodiment in the Arts and Culture*, Berkeley: University of California Press, 2009.

"Early Tape Recordings," *Audio Engineering Society*, Audio Engineering Society, 29 December 2002. <www.aes.org> Accessed 3 August 2016.

"Echoes & News: The Musical Medium, Aubert," *The Annals of Psychical Science*, London: Office of the Annals, 1905.

Edgerton, Gary R. *The Columbia History of American Television*, New York: Columbia University Press, 2007.

Edison, Thomas A. and Dagobert D. Runes. *The Diary and Sundry Observations of Thomas Alva Edison*, New York: Philosophical Library, 1948.

Ehman, Jerry. "'Wow!'—A Tantalizing Candidate," in H.P. Shuch, *Searching for Extraterrestrial Intelligence: Seti Past, Present, and Future*, Berlin: Springer, 2011.

Elliman, Paul. "Detroit as Refrain," *The Serving Library,* Bulletins of The Serving Library, no. 8, 2014. <www.servinglibrary.org/journal/8/detroit-as-refrain> Accessed 5 March 2016.

Eveleth, Rose. "Earth's Quietest Place Will Drive You Crazy in 45 Minutes," *Smithsonian Magazine*, December 17, 2013. <www.smithsonianmag.com/smart-news/earths-quietest-place-will-drive-you-crazy-in-45-minutes-180948160/?no-ist> Accessed 23 January 2016.

Feaster, Patrick. *Pictures of Sound: One Thousand Years of Educed Audio: 980-1980,* Dust to Digital, 2012.

Finucane, Ronald C. *Appearances of the Dead: A Cultural History of Ghosts,* Buffalo, NY: Prometheus Books, 1984.

"First Motion Pictures Transmitted by Radio Are Shown in Capital," *Washington Post,* Washington, DC: June 14, 1925.

Fischer, Andreas. *Okkulte Stimmen—Mediale Musik: Recordings of Unseen Intelligences 1905-2007,* Berlin: Supposé, 2007. CD.

Fisher, Mark. *Ghosts of My Life: Writings on Depression, Hauntology and Lost Futures,* Winchester, UK: Zero Books, 2014.

"FM Signals Sent as 'Radio Pill' Passes through Body," *Electrical Engineering,* 76.6, New York: American Institute of Electrical Engineers, 1957.

"From Whence the Moog?" Dir. Unknown. Perfs. Herbert Deutsch. 15 January 1983. Unreleased video footage transferred to DVD, The Henry Ford Museum, 2014.

Galbraith, Robert M. to George E. Howard, November 10, 1929. Edison Papers. Unprocessed. Benson Ford Research Center, The Henry Ford, Dearborn, Michigan. Personal Correspondence.

Garrett, C.B.G. "The Optical Maser," *Electrical Engineering,* 80.4, 1961.

Gertner, Jon. *The Idea Factory: Bell Labs and the Great Age of American Innovation,* New York: Penguin Press, 2012.

"Ghosts," *Joy Division,* Dir. Gee, Grant. Madrid: Avalon, 2008. Film.

Glassie, Henry. *Vernacular Architecture,* Philadelphia: Material Culture, 2000.

Glinsky, Albert. *Theremin: Ether Music and Espionage,* Urbana: University of Illinois Press, 2000.

Grierson, Francis. *Psycho-phone Messages,* Los Angeles, Calif: Austin Publishing company, 1921.

Guins, Raiford. *Game After: A Cultural Study of Video Game Afterlife,* MIT Press, 2014.

Hammer, Peter. "In Memoriam," *Journal of the Audio Engineering Society,* Audio Engineering Society, Vol. 42, No. 6, June 1994.

"Haunted Places in Wyoming," *The Shadowlands,* 1998. <www.theshadowlands.net> Accessed 3 August 2016.

Helmreich, Stefan. "An Anthropologist Underwater: Immersive Soundscapes, Submarine Cyborgs, and Transductive Ethnography," in Jonathan Sterne, *The Sound Studies Reader,* New York: Routledge, 2012.

Heys, Toby. "Re: Spurious Frequencies," Message to Kristen Gallerneaux, 1 October 2015. Email.

Hilderbrand, Lucas. *Inherent Vice: Bootleg Histories of Videotape and Copyright,* Durham: Duke University Press, 2009.

"History of the Detroit Salt Mine," *Detroit Salt Co.*, <detroitsalt.com/history/> Accessed 28 January 2016.

Hultén, Karl Gunnar Pontus. *The Machine as seen at the End of the Mechanical Age*, New York: The Museum of Modern Art, 1968.

Jenkins, C.F., *The Boyhood of an Inventor*, Washington, DC: Printed by National Capital Press, Inc., 1931.

Jennings, David. *Skinflicks: The Inside Story of the X-Rated Video Industry*, Bloomington, IN: Distributed by 1st Books Library, 2000.

John Robinson Pierce: Music from Mathematics, Volume Two, Finders Keepers Records / Cacophonic: CACK4504, 2013.

Jones, David E.H. *The Inventions of Daedalus: A Compendium of Plausible Schemes*, W.H. Freeman & Company, 1982.

Joy Division. Dir. Gee, Grant. Madrid: Avalon, 2008. Film.

Kaprow, Allan. *Essays on the Blurring of Art and Life*, Berkeley: University of California Press, 1993.

Kittler, Friedrich. *Gramophone, Film, Typewriter*, Stanford, Calif: Stanford University Press, 1999.

Knittel, Chris. "The Mystery of the Creepiest Television Hack," *Motherboard*, Vice Media LLC, 25 November 2013 <motherboard.vice.com/read/headroom-hacker> Accessed 31 October 2017.

Lanza, Joseph. *Elevator Music: A Surreal History of Muzak, Easy-Listening, and Other Moodsong*, New York: St. Martin's Press, 1994.

Larson, T A. *History of Wyoming*, Lincoln: University of Nebraska Press, 1966.

Laser Light; a New Visual Art: Organized by the Cincinnati Art Museum and the Laser Laboratory of the University of Cincinnati Medical Center, Cincinnati Art Museum, November 12 Through December 14, 1969, Cincinnati Art Museum, 1969.

"Laser Pointer Irks Kiss," *Beaver County Times,* Beaver County: PA: McClatchy-Tribune Information Services, November 24, 1998.

Law, John. *After Method: Mess in Social Science Research*, London: Routledge, 2004.

Lear, John. "Now, the Broadcasting Pill," *New Scientist*, London: New Science Publications, 18 April 1957.

Lethbridge, Thomas C. *Ghost and Ghoul*, London: Routledge & Kegan Paul, 1961.

Lesy, Michael, Schaick C. Van, and Warren Susman. *Wisconsin Death Trip*, New York: Pantheon Books, 1973.

Littler, Dr. R. "'We Watch You While You Sleep' TV signal intrusion 1975," *Scarfolk Council*, Richard Littler. 19 February 2014 <www.scarfolk.blogspot.com/2014_02_01_archive.html> Accessed 30 March 2016.

Loomis, C.G. "Illinois Poltergeist." *Western Folklore*. 17.1, 1958.

Lubachez, Ben J. *The American Cinematographer*, "The Beginnings of the Cinema: Birth of the Final Form of the Motion Pictures—The Work of C. Francis Jenkins," Hollywood, Calif: ASC Holding Corp, May 1, 1922.

"Magdalena Parker," AUDINT. <www.audint.bigcartel.com/product/magdalena-parker> Accessed 28 January 2016.

"Magnetic Recording History Pictures," *Audio Engineering Society*, Audio Engineering Society, 6 July 2004. <www.aes.org> Accessed 3 August 2016.

"Making the Perfect Searchlight," *New Scientist*, London: New Science Publications, May 12, 1960.

"Mars' Latest Visit Leaves Riddle of Life Unsolved: Red Planet Probed by Watchers in Search for Clues That Life Exists on Its Surface," *Popular Mechanics Magazine*, New York: Hearst Corp., January 1927.

"Max Sets Bad Speech Habit," *Bryan Times*, Bryan, Ohio, 10 August 1987.

McLellan, Dennis. "Composer was a synthesizer pioneer," *Los Angeles Times*, 11 January 2008 <www.articles.latimes.com/2008/jan/11/local/me-garson11> Accessed 6 April 2016.

McLuhan, Marshall. *Understanding Media: The Extensions of Man*, New York: McGraw-Hill, 1964.

Moberly, C.A.E, Eleanor F. Jourdain, Edith Olivier, and J.W. Dunne. *An Adventure*, London: Faber and Faber, 1931.

"Montezuma Mansion," *Weird California* <www.weirdca.com/location.php?location=64> Accessed 30 January 2016.

"Moog Prototype Unsold at Auction," *Keyboard*, San Francisco, CA: Miller Freeman Publications, September 1982.

"Moving Pictures Sent by Radio," *Popular Mechanics*, Chicago, Ill: Popular Mechanics Co, September 1925.

"Mount Vernon Arts Lab – The Séance at Hobs Lane – Description," *GhostBox*, GhostBox, 2001 <ghostbox.greedbag.com/buy/sance-at-hobs-lane-0/> Accessed 6 April 2016.

"Music and the Paranormal," Melvyn Willin <www.hulford.co.uk/musicpa2.html> Accessed 30 January 2016.

National Capital Park and Planning Commission, *Maryland Historic Sites Survey (M:31-10)*, "Jenkins Broadcasting Station," Maryland Historical Trust, August 21, 1975.

Novak, Colin. "Summary of the Windsor Hum Study Results," *Global Affairs Canada*, Government of Canada, 23 May 2014 <www.international.gc.ca> Accessed 26 January 2016.

"Organ Music," *San Diego Paranormal Research*. n.d. <www.sdparanormal.com/Villa_Montezuma.html> Accessed 30 January 2016.

Paglen, Trevor, and Rebecca Solnit. *Invisible: Covert Operations and Classified Landscapes*, New York, NY: Aperture, 2010.

Parikka, Jussi. *What Is Media Archaeology?* Cambridge, UK: Polity Press, 2012.

Pasewalk, L.P. to Henry Ford. BFRC.REG.R229.File 31.884, Benson Ford Research Center, The Henry Ford, Dearborn, Michigan. Personal Correspondence.

Phillips, Tony. "The Tunguska Impact—100 Years Later," NASA Science: *Science News*, NASA, 2008 <science.nasa.gov/science-news/science-at-nasa/2008/30jun_tunguska/> Accessed 28 January 2016.

Phipps, Tony. "Can Computers Act?" *Screen Actor* Vol. 22-28, 1980, Los Angeles, Calif: Screen Actors' Guild.

Pinch, Trevor J. and Frank Trocco. *Analog Days: The Invention and Impact of the Moog Synthesizer*, Cambridge, Mass: Harvard Univ. Press, 2004.

Pourade, Richard F. *The Glory Years*, San Diego: Union-Tribune Pub. Co., 1964.

Radio Pill. British Pathé, 1961 <www.britishpathe.com//pages/in-the-news/> Accessed 6 March 2016.

"Radio Vision Shown First Time in History by Capital Inventor," *Sunday Star,* Washington, DC: June 14, 1925.

Reminiscence, "Written Reminiscence of the Piccard Stratosphere Balloon Flight of October 23, 1934," ARCH.13, Box 18, Folder: "Piccard Stratosphere Balloon Flight," n.d., Benson Ford Research Center, The Henry Ford, Dearborn, Michigan.

Reynolds, Simon. *Retromania: Pop Culture's Addiction to Its Own Past*, London: Faber & Faber, 2011.

Riordan, Michael. "How Bell Labs Missed the Microchip," *IEEE Spectrum*, New York: Institute of Electrical and Electronics Engineers, 1 December 2006 <www.spectrum.ieee.org/computing/hardware/how-bell-labs-missed-the-microchip>. Accessed 23 January 2016.

Rogo, D. Scott. *Nad: A Study of Some Unusual "other-World" Experiences*, New York: University Books, 1970.

Rosemary's Baby. Dir. Polanski, Roman. Paramount Pictures, St. Louis, Missouri: Swank Motion Pictures, 1968.

Sconce, Jeffrey. *Haunted Media: Electronic Presence from Telegraphy to Television*, Durham, NC: Duke University Press, 2000.

Seafield, Lily and Rosalind Patrick. *Ghostly Scotland: The Supernatural and Unexplained*, New York: Barnes & Noble, 2006.

Scandura, Jani. *Down in the Dumps: Place, Modernity, American Depression*, Durham: Duke University Press, 2008.

Schulman, J. Neil. Appendix: "Violet: House of the Laser," in *The Rainbow Cadenza: A Novel in Logosata Form*, New York: Simon and Schuster, 1983.

Schwartz, R.N. and C.H. Townes. "Interstellar and Interplanetary Communication by Optical Masers," Institute for Defense Analyses, *Nature: Journal of Science*, London: Macmillan Journals Ltd, April 15, 1961.

"Sonic Disruptions," Tate Modern <www.tate.org.uk/whats-on/tate-britain/talks-and-lectures/sonic-disruptions> Accessed 28 January 2016.

Sorenson, Paul. "Looking Back..." *Aerospace Engineering and Mechanics Update*, Minneapolis: University of Minnesota Institute of Technology, 1998–99.

Spence, Lewis, Leslie Shepard, and Nandor Fodor. *An Encyclopedia of Occultism: A Compendium of Information on the the Occult Sciences, Occult Personalities, Psychic Science, Magic, Demonology, Spiritism, Mysticism and Metaphysics*, New York: University Books, 1960.

"Strange Instrument Built to Solve Mystery of Cosmic Rays," *Popular Science*, New York: Popular Science Pub. Co., April 1932.

Tandy, Vic & Tony R. Lawrence. "The Ghost in the Machine," *Journal of the Society for Psychical Research* 62, no. 851, April 1998.

Taussig, Michael T. *My Cocaine Museum*, Chicago: University of Chicago Press, 2004.

Telegram from the Secretary of the Navy to All Naval Stations Regarding Mars, August 22, 1924, Record Group 181, Records of Naval Districts and Shore Establishments, 1784-2000, ARC Identifier 596070, National Archives and Records Administration.

"The Advisory Circle—From Out Here—Description." *GhostBox*. GhostBox, 2015 <ghostbox.greedbag.com/buy/from-out-here/> Accessed 6 April 2016.

The Henry Ford Museum, Accession File #63.173.1.

"The Historic 12 December Event: The Year 1960," *The History of the Laser*, 2012 <www.alijavan.mit.edu/Original/Historic/12December1960event.htm> Accessed 23 January 2016.

The Jack Mullin Collection, *Pavek Museum,* Pavek Museum of Broadcasting, 2016 <www.pavekmuseum.org> Accessed 3 August 2016.

The Museum of Classic Chicago Television <ww.fuzzymemories.tv/> Accessed 30 March 2016.

"The New Edison: The Phonograph with a Soul," *The Paris Morning News*, Paris, Texas, 13 February 1921.

"The Psycho Phone," *The New Yorker*, New York: New Yorker Magazine, inc., etc., 1933: Vol. 9.

"The Quenching of Sayville: The Close of a Sordid Story," *The Wireless World* 3, no. 32, November 1915.

The San Diego Union, San Diego, Calif: Union-Tribune Pub. Co., July 20, 1913.

"The Villa Montezuma: History of a House." *Friends of the Villa Montezuma.* n.d. <www.villamontezumamuseum.org/Villa_Site/About_us-Main_pg.html> Accessed 30 January 2016.

Thompson, Emily A. *The Soundscape of Modernity: Architectural Acoustics and the Culture of Listening in America, 1900-1933,* Cambridge, Mass: MIT Press, 2002.

Thompson, Stith. *The Motif-Index of Folk-Literature*, Bloomington: Indiana University Press, 1955–58.

Tilley, Christopher and Wayne Bennett. *The Materiality of Stone: Vol. 1,* Oxford: Berg, 2004.

Tompkins, Dave. *How to Wreck a Nice Beach: The Vocoder from World War II to Hip-Hop, the Machine Speaks*, Chicago, IL/Brooklyn, NY: Stop Smiling Books/Melville House Publishing, 2010.

Todd, Mabel Loomis. *Total Eclipses of the Sun*, Boston: Roberts Bros., 1894.

Vent, Bonnie. "Jesse Shepard Channel: 12/07/2002," *San Diego Paranormal Research*. n.d.
<www.sdparanormal.com/articles/article/274216/1670.htm> Accessed 30 January 2016.

Volsupa. "WE RUN THIS." Online video clip. *YouTube*, YouTube. Uploaded on 11 September 2008. <www.youtube.com/watch?v=j0ZFow_9vsg> Accessed 30 March 2016.

"Weird 'Radio Signal' Film Deepens Mystery of Mars: Pictorially Recorded Messages Here Mere Tangles Mass of Dots and Dashes—Growing Wonderment May Bring Tenable Interpretation Theory," *Washington Post*, August 27, 1924.

Wells, Joseph Nicholas. "Siltpile at Scottsdale," *Journal of Architectural Education, 1947–1974,* Vol. 18, no. 3 (December 1963).

"WFLD Channel 32 – 'Pirate Report' (1987)" *The Museum of Classic Chicago Television*. The Museum of Classic Chicago Television. Original air date, November 23, 1987. <www.fuzzymemories.tv/index.php?c=59&m=max%20headroom%20pirate#videoclip-2465> Accessed 30 March 2016.

When the Wind Blows. Dir. Murakami, Jimmy T. Writer. Raymond Briggs. New York, N.Y: IVE, 1986. Film.

"WLS Channel 7—'Pirate Report' (1987)." *The Museum of Classic Chicago Television*. The Museum of Classic Chicago Television. Original air date, November 23, 1987. <www.fuzzymemories.tv/index.php?c=59&m=max%20headroom%20pirate#videoclip-2468>. Accessed 29 March 2016.

"WMAQ Channel 5 – 'Pirate Report' (1987)" *The Museum of Classic Chicago Television*. The Museum of Classic Chicago Television. Original air date, November 23, 1987. <www.fuzzymemories.tv/index.php?c=59&m=max%20headroom%20pirate#videoclip-2466> Accessed 30 March 2016.

Wojcik, Daniel. *The End of the World As We Know It: Faith, Fatalism, and Apocalypse in America*, New York: New York University Press, 1997.

Zoglin, Richard. "Grounding Captain Midnight: High-tech, Holmesian Detective Work Unmasks a Satellite Intruder," *TIME*, 4 August 1986.

INDEX

AEG (Allgemeine Elektricitäts-Gesellschaft), 180, 201
'Affordances', 12
American Civil War, *xiv*, 57
American Society for Psychical Research, 29, 49, n.16
American Telegraphone Company, 179
Amherst College, 67, 72 n.22
Ampex Corporation, 181-83
An Adventure (Moberly and Jourdain), 32
Anthropocene, 200
Apple, Inc., 212
Architecture, 5, 9-10, 44, 117, 165, 187, 195, 212
Archives, *xiv*, *xv*, 29, 129, 167, 260, 264
Ashtar Galactic Command, 158, 174 n.4
Astronomy, 67, 119, 235
Atari, 130, 263
Atlantic Communication Company, 97, 100
AUDINT, 6, 193-98, 201-7, 263
Aurora borealis, 200, 263
Baldoon Mystery, 228
BASF, 180, 201

Belief, paranormal culture, and folklore:
Anomalous events, 40, 48, 156
Apocryphal stories, 122 n.3, 234
Apparitions, 10, 141, 143, 170, 186-90, 199, 219, 253,
Apportation, *xiii*, 53-54, 227
Belief, x, 5, 13, 15, 28, 30, 38-40, 43, 47, 48, 70
Channeling, 11, 14, 25-48, 50 n.32, 50 n.45, 62, 172, 190, 202
Clairaudience, 155
Contact magic, 40
'Continuity and change' (folkloric process), 33
Crisis apparitions, 141, 187
Ectoplasm, 114, 259
EVP (electronic voice phenomena), 10-11, 186
Family lore, *x*
Fata morgana, 62
Fetch, 139-48, 188
Folklore, folkloric, *x*, 5, 19, 33, 123 n.14, 143, 169, 193, 201, 228, 236, 250, 260-63
Frauds, charlatans, 25-48, 49 n.16, 114, 158
Ghost hunters, 11, 46
Ghost ships, 62, 136, 258, 262
Ghosts, 5-6, 9-10, 14-16, 20-22, 32, 33, 44-45, 139-48, 152, 169-73, 185-91,

197, 200, 236-37, 241-43, 252, 255 n.33, 257, 259, and passim; and haunting, 4-7, 10-11,13, 25-48, 71, 103, 139-49, 155-74, 185-91, 199-200, 204, 220, 227-29, 236-37, 248, 255, and passim; see also hauntology

Legends, *x*, 33, 34, 40, 44, 50 n.45, 142, 196, 237, 251, 253-64; legend-tripping, 249

Liminal, liminality, 43, 47, 172, 189, 220, 250

Lithobolia, 13

Manifestations, 4, 6, 14, 15, 30, 34, 36, 40, 42, 45, 114, 186, 208 n.8, 214, 227, 251

Mars, martians, extraterrestrials, UFOs, *xiv*, 6, 63, 64-71, 85, 119-20, 124 n.30, 124 n.34, 174 n.5, 199, 242, 243, 258, 259

Mediums, 25-47, 49 n.16, 190

Memorates, 33

Moberly-Jourdain incident, 32

Musical séances, 5, 34-5, 42-3

Numinous experiences, 33, 40

Ouija boards, 80, 187

Para-acoustic sound, 47

Paranormal culture, 5, 15, 260

Paranormal music, 30-33, 50 n.44

Parapsychology, psychic research, *xiii*, 9-11, 14-16, 27, 29, 30, 31, 35, 199, 262

Phone calls from the dead, 25-29, 36, 45, 103, 141-42, 145, 147, 258

Poltergeists, 13-15,32, 47, 227-29

Possession, 30, 165, 206

Premature burial, *x*, 4, 30

Psychic photography, thoughtography, *xi-xii*

Residual, residue, imprints (spiritual), *xiv*, 1, 9-16, 26, 32, 47, 187, 204, 229, 248, 262

Rituals, 11, 20, 22, 43, 64, 203, 205, 217-224

Scientology, 164, 172

Skepticism, 33, 64, 70, 259

Spiritualist, spiritualism, séances, *x*, *xiv*, 5, 6, 14, 25-48, 63, 114, 172, 186, 190, 257-58, 260-63

Stone tape theory, 9-11, 13, 15, 133, 237

Synesthesia, 222

Theosophy, 36

Wow! Signal, 159, 174 n.5

Bell Telephone Laboratories, 105, 108, 110, 112, 113, 120, 121, 122 n.10, 123 n.15, 128, 203, 204, 209 n.24

Breakthrough (Raudive), 11, 186
Caltech, 116-17, 123 n.23
Canadian National Railway, 249
Carnegie Institute of Technology, 79
Chelyabinsk meteor impact, 200-201
Central Intelligence Agency (CIA), 197, 258
Ciphers, codes, censors, 29, 60, 63, 67, 80, 81, 98-101, 145, 179, 194, 197, 213, 237, 238 n.9
Cold War, 69, 108
Curators, 7
Curiosity Rover, 119-20
Cybernetics, 116
Declaration of Neutrality, 98
Depression Era, 28, 65
Digital Equipment Corporation, 132-33
Domino's Pizza, 131
Dreamworlds of Alabama, xiii
Dryden Community Pictorial Quilt, 82
Dryden Ladies Library Association, 82
Dungeons & Dragons, 254
eBay, *xi*, 28, 210
Experiments in Art and Technology, 115
Face on Mars, 120

Famous Monsters of Filmland, xii, xiii
Federal Bureau of Investigation (FBI), 88, 163, 167
Federal Screw Works, 128
Film and video:
Burnt Offerings, x
C.H.U.D., xii
Caltiki the Immortal Monster, xii
Cat-Women of the Moon, 261
Child's Play, 252
Close Encounters of the Third Kind, 159
Clutch Cargo, 161
Crystals Go To War, 204
Dawn of the Dead, xiii
Deep Red, 221
Delusions of the Living Dead, 196, 197, 203
Dr. Who, 'The Horror of Fang Rock', 158-160
Evil Dead, The, 151-2
Fog, The, 136
From Whence the Moog?, 223
Gingerdead Man, The, xiii
Golden Voyage of Sinbad, The, xii
Halloween, 221
Incredible Melting Man, The, xi-xii, xiv
Invocation of My Demon Brother, 221

Manitou, The, 260
Merrie Melodies, 'Falling Hare', 159-60
Morel's Invention, 11
Night of the Living Dead, 221
Principles of the Optical Maser, 106-7, 109, 112-13, 121 n.1
Rosemary's Baby, 170
Shining, The, 264
Star Trek, 88
Star Wars, 159
Suspiria, 221
Thing, The, 259
Twilight Zone, The, 160
Unsolved Mysteries, 259
Warriors, The, 132
When the Wind Blows, 157
Wicker Man, The, xii
Forests, 6, 88, 186, 201, 237, 248, 250, 253, 254, 255
From India to the Planet Mars (Flournoy), 63
Gambols with Ghosts, 34
Geological samples:
  brick, 11; crystal, *x*, 7, 146, 204-5, 209 n.24; galena, 7; granite, 11; limestone, 11, 204, 259; quartz, 9-10, 11, 204; rocks, 11, 12, 120, 159, 202, 204; stone, *x*, 9, 10, 13, 15, 57, 107, 139-40, 145, 227

Ghost Army, 194
Ghost Box Records, 221
Gothic Futurism, 111
Graveyards, gravesites, cemeteries, *x*, 30, 195, 253, 255
Griffith Observatory, 117
Hackers, hacking, 164-65, 168, 175 n.9, 175 n.12
Hagley Museum, 95
Harvard Sentences, 129
Haunted Media, 49 n.13, 169
Hauntology, 5-6, 170-74, 197, 208 n.7, 220
Heathkit, 130
Henry Ford Museum, *xiii-xv*, 6, 65, 70, 95, 120, 223, 233, 261, 264
Hidden rooms, 39, 146, 185
Hofstra University, 217, 223
Hughes Aircraft Company, 108
International Business Machines Corporation (IBM), 123 n.19
Infrared, 105, 106, 119, 120
InnerConn Technology, 242
Installation art, 116, 123 n.26, 197, 205
Institut für Grenzgebiete der Psychologie, 30
Institut Général Psychologique, Paris, 35
Institute of Electrical and Electronics Engineers, 120

Institute of Radio Engineers, 181
IREX, 206
James Hislop Co., 79
Johns Hopkins University, 80
Kmart, *xiii*
Korean War, 108
Laboratories, 15, 16, 57, 60, 61, 65, 67, 69, 80-81, 105, 106, 110, 112, 120, 121 n.1, 123 n.15, 124 n.34, 128, 131, 199, 203, 204
Lasik procedure, 118
Lebedev Physical Institute, 108
Libbey Glass Co., 75-6
Light (publication), 40-41
Light, coherent, 107-8, 111, 118; incoherent, 112
Lusitania, 98-99
Magic of the Horseshoe, The, 261
Masks, *xiv*, 159, 163-69, 173, 174 n.2, 259
Material culture, material history, x, 4, 13, 14, 15, 106, 203, 204, 207, 227
Max Headroom Incident, 158-174
Medical conditions, hearing disorders:
 Meniere's Disease, 155-56, 173, 257; Moebius syndrome, 131; Tinnitus, 156

Menlo Park, New Jersey, 26, 235, 236
Michigan, University of, 121, 131; Artificial Language Laboratory, 131
Military and naval technology, 59, 60, 61, 64, 66, 67, 70, 80, 108, 113, 113, 114, 179, 181, 199, 203, 207
Mills Folklore Archives, 260
Mines and miners, 195-6, 204, 207 n.1, 236, 237
Mood Music (Bingham), 79
Museum of Modern Art (MoMA), 116
Museums, *xiv*, 6, 7, 9, 11, 30, 44, 45, 65, 70, 95, 116, 120, 121, 166, 167, 201, 211, 223, 233, 261

Music:
 Advisory Circle, *From Out Here*, 218
 AUDINT, *Martial Hauntology*, 197
 Black Sabbath, 'Iron Man', 3
 Carlisle, Belinda, 'Heaven Is A Place On Earth', 157
 Cybotron, *xii*
 Deutsch, Herb, 'Jazz Images: A Worksong and Blues', 219
 Emerson Lake & Palmer, 117

Fat Boys, 262
Four Tops, The, 'I'll Turn To Stone', 1
Garson, Mort, *Ataraxia: The Unexplained*, 220; *Black Mass Lucifer*, 220; with Paul Beaver, *The Zodiac: Cosmic Sounds*, 220
Gaye, Marvin, 'What's Going On', 262
Goblin, 221
Joy Division, 171
Jr. Walker & the All Stars, 'Shotgun', 207
Kiss, 'Beth', 114
Kraftwerk. 'Uranium', 127
Mount Vernon Arts Lab, *Séance at Hobs Lane*, 222
Pink Floyd, *Dark Side of the Moon*, 117
Public Enemy, 'Welcome To The Terrordome', 169
Rolling Stones, 117
Tiffany, 'I Think We're Alone Now', 157
Wham!, 262
Williams, Hank, 'Your Cheatin' Heart', 2

Musical instruments:
Drums, *xiii*, 32, 42
Piano, *xiv*, *xiv*, 2, 32, 34, 36, 39, 41, 42, 43, 46, 222, 265
Synthesizers: Abominatron, 219; Dirt synth, 227-29; Minimoog Model D synthesizer, 221; Modular synthesizers, modules, 220-24; Moog synthesizer prototype, *xiv*, 217-24; MoogIIIP, 221
Theremin, 94, 217
Turntables, 63, 202

National Electric Supply Co., 64
New Scientist, 93, 95, 119
New York Herald, 235
Native America: Algonquin, 136, 260; Hopi, 59; Huron, 260; Inuit, 263; Metis, 260; Ojibwa, 260
OMNI, *xiii*
Pace Gallery, 115
Patents, *x*, 28, 58, 60, 81, 83 n.4, 83 n.5, 100, 107, 118, 122 n.9, 124 n.27, 124 n.28, 124 n.29, 178 n.3, 201, 208 n.15, 261
"Paul Is Dead" rumor, 62
Pennsylvania, University of, 235
Pepsi Pavilion, 115
Petit Trianon, 32
Petrozavodsk phenomenon, 159
Pictorial History of Horror Movies, A (Gifford), *xiii*

Pirates, piracy, 158-76
Pontiac Silverdome, 262
Popular Radio, 83
Principles of Scientific Management, The, 82
Prospectors stake, 6, 233-37
Prototypes, *xiv*, 58, 60, 93, 128, 181, 212, 217-24
Psycho Phone Messages (Grierson), 25-27, 36
QST magazine, 87
Radio and television:
  networks: ABC, 181; BBC, 10, 180; CBS, 161, 221; Christian Broadcasting Network, 168; Granada TV, 170; HBO, 167; ITV, 159; Jenkins Radiomovie Broadcast Station (W3XK), 58; Nauen wireless station, 81, 98, 100; shows: Dr. Demento, 157, Night Flight, 157
Radioshack, 130, 212
Radio Corporation of America (RCA), 61, 93, 94
Reagan era, 157
Red Lobster (restaurant), *xi-xii*
Reeves Sound Labs, 204
"Religious Takeover, The", 168
Rockefeller Institute, 93
Saducismus Triumphatus (Glanvil), 32
Salt, salt mines, *x*, 11, 145, 187, 193-95, 201, 204, 205, 207, 207 n.1, 207 n.6, 211 n.7, 261, 263
Samuel Oschin Planetarium, 117
Scarfolk Council, 170
Science fiction, 106
Sea and oceans, 57-62, 63, 88, 98, 100, 109, 110
Sears Tower, 166
SETI Institute, 70, 119, 124 n.30, 159, 174
Signal Intelligence Service, 67
Silk Road, 206
Smithsonian, 129, 223
SMPTE, 129, 157, 162
Solar eclipse, 233, 234

Sound phenomena:
  Acousmatic sound, 31, 173
  Acoustics, 5, 11, 25, 26, 30, 40, 42-3, 45, 47, 48, 128-9, 156, 196, 197, 242
  Archaeoacoustics, 9
  Bass sounds, bass music, 42, 156, 186, 198, 199, 201, 263
  Easy-listening, 82
  Echo, 5, 6, 10, 39, 42, 46, 107-8, 167, 169, 202, 220, 221, 229
  Educed audio, 9, 69

Female voice, 130, 132-34
Formants, phonemes, xiv, 127-29, 133
Frequencies, 6, 66, 69-70, 93-95, 97-101, 107, 109, 112, 127-33, 186, 193-207, 214, 235, 263
Glitches, 66, 169
Heterodyning, 7, 63, 202-3
Hum, The, 199-200, 263
Infrasound, 198-202, 228, 263
Klaxon horns, 156
Laugh tracks, 4
Mimesis, mimicry, 26, 33, 42, 43, 60, 63, 128, 163, 169, 199
Ohrwurm, ear worm, 197, 200, 203, 207 n.8
Oscillations, 13, 93-4, 116, 121 n.1, 173, 204, 205-7, 218-20, 245
Paleospectrophony, 69
Resonance, 7, 9, 29, 151, 196, 229
Reverb, xiii, 172, 189, 200, 243
Ring modulation, 205
Signals, transmissions, passim
Sound waves, radio waves, radio signals, 4 and passim
Speech, 26, 43, 110, 111, 128-9, 131-33, 164, 165
Static, 1, 4, 61, 63, 68-9, 71, 99, 110, 113, 172, 181, 186, 188, 221, 254, 258, 263
Sub-frequencies, 186
Test-tones, 103, 156, 207
Unsound, 194, 197
Vibrations, ix, 9, 12, 94, 115, 116, 196, 204
VLF waves, 200, 263
Voice, 3-5, and passim, see also speech
Voltage control, 219, 220
Southern Pacific Railroad, 241
Southern Television Broadcast Interruption, 158-60, 167, 170
Space exploration, 64, 69, 70, 86, 123, 159
Stevens Institute of Technology, 235
Stimulus progression, 82
Stollwerck Brothers, 97
Stratosphere balloon, 85-8, 119
Studies in Melody (Bingham), 79
Submarines, 98, 100
Tape music, 116, 205-6, 217, 219
Technology:
 "The Thing" (espionage device), 94
 Active Denial System, 113
 Ampliphone, 99

Antenna, 6, 27, 67, 81, 87, 94, 186, 197, 213

ARPAnet, Internet, 81, 111, 120, 147, 159, 167, 172, 179

Black boxes, 7, 65,

Computers: Control Data 6500, 131; iMac G3, 212; PDP-8, 133, 175 n.9; punch cards, 253; TRS-80, 130

Doorbells, 218

Fiber optics, 110-11

Film and cinema, *x*, *xi*, 7, 11, 13, 59, 60, 63, 67, 68, 94, 109, 112, 116-17, 120, 132, 152, 170, 196, 200, 204, 205, 221, 252, 260, 262; Phantoscope motion picture projector, 58

Incandescent lamp, lightbulb, 53, 112, 185, 190, 234, 235, 236, 257

iPod, 213

Lasers and masers, *x*, 5, 105-21, 122 n.3, 122 n.6, 122 n.9, 122 n.10, 128, 189, 260; laser art, 114-18, laserium, 117; laser pointers, 113-14, 117, 118; Laser Weapon System (LaWS), 113

Microphones, 6, 132, 156, 187, 197, 205, 228, 261

Microwaves, 105, 106-9, 113, 119, 121 n.2, 122 n.4, 166

Photography, 7, 13, 44, 57, 66, 85, 86, 112, 147, 152, 170, 198, 207, 246, 263; psychic, *xi*

Radio, wireless, *xiv*, 1, 6, 7, 13, 29, 30, 58-62, 62-71, 80, 81, 83, 86-88, 97-101, 107, 109, 111, 119, 132, 157, 174, 180-92, 186, 190, 194, 203-205, 261, 262, 263; radio control, 262; radio facsimile, 60; radio pill, *x*, 5, 93-95; and prison, 211-14; radios: SE 950, 65, 69; Sony SRF-39FP, 211-15; Western Electric SCR-67 and 68, 81

Recorded sound: cassette tape, 156, 205-6; Dictaphone, 181; Ediphone, 180; magnetic recording, 11, 116, 166-7, 179-80, 187, 201; Muzak, xiii, 82-3; phonographs, 4, 13, 26, 35, 41, 79, 99, 100, 180, 197, 204; reel-to-reel recorder, 11, 201, 219, 263; telegraphone, 179; wax-cylinders and cylinder recordings, 4, 99-100, 180; wire recording, 179-80

Robots, HERO 1, 130-31

Tasimeter, 234
Telegraphy, 13, 27-9, 61, 98, 99, 109-10; and Morse code, 29, 80, 81, 98
Telemetry, 95
Telephones, 4, 13, 28, 29, 30, 103, 110, 141, 142, 175 n.9, 176 n.19, 197, 212-13; cell phones, 6, 188, 253, 257; telephony by trees, 81
Television, 5, 6, 7, 10, 11, 57, 58, 60, 61, 62, 65, 94, 95, 115, 129, 130, 155-76, 181, 212, 214, 221; Iconoscope, 95; Jenkins Radio Camera, 66-7; and piracy, 158-76; Radiovisor, 65; sign-offs, 157, 173
Text-to-speech, 128, 164; DECtalk, 136-37; Speech Plus, 137; synthesis, 136-37; Votrax, SC-01 chip, *xiv*, 127-30; Votrax, Type 'N Talk, 128-9
Transistors, 93, 95, 120-21, 128, 182, 212-13, 219
VHS tape, *xii*, 166, 254
Vocoder, 261
Walkman, 211, 247
Weapons, 108, 113, 114, 159
YouTube, 167, 170
Telefunken, 97-8, 261
Thresholds, 2, 11, 15, 156, 173, 186, 189, 251
Tropication Arts, 229
United States Life Saving Service, 58-9
United States Government, 85, 97, 98, 163, 193, 199, 207
United States Naval Observatory, 67
United States Secret Service, 99, 100, 163
United States Signal Corp, 180-84
Valley of Shadows (Grierson), 36
Veteran's Administration Hospital, 93
Video Games:
 Black Hole, 130; *E.T.—The Extra-Terrestrial*, 264; *Gorf*, 130; *Q*Bert*, *xv*, 130; *Reaktor*, 130; *Wizard of Wor*; 130
Xenon, 108
Vietnam, *xii*, 206
Villa Montezuma, 35-45
Vocal Interface Division, 128
Weather, 57-62, 69, 75, 99, 110, 214
Wisconsin Death Trip, *xiii*
World, The (periodical), 46
World Exposition of Paris, 179
World of Ted Serios, The, *xiii*
WWI, 80, 81, 98, 101, 200

WWII, 95, 165, 180-181, 194, 200, 201, 204
Yorkshire Spiritual Telegraph, The, 28
Zombies, *xiii*, 132, 156, 221

NAMES

Abbott, Martin H., 118
Amiss, Kevin T., 118
Anger, Kenneth, 221
Anonymous (hacktivist group), 164-65, 172
Anthes, John, 116
Apgar, Charles, 99-101
Arendt, Hannah, 70
Argento, Dario, 221
Armat, Thomas, 58
Arnett, Bill, 194, 202

Bacci, Marcello, 30
Bacon, Roger, 220
Banim, John, 135
Barker, George, 235
Barrett, William, 10
Basov, Nikolay, 108
Beaver, Paul, 220
Beecher, Henry Ward, 37
Bennett, Wayne, 12 n.11
Bennett, William, 105, 108
Bingham, Walter Van Dyke, 79-80
Blavatsky, Helena, 36
Bozzano, Ernest, 31
"bpoag", 168
Briggs, Raymond, 152-53
Britten, Emma Hardinge, 29, 34
Brown, Bill, 12
Brown, Danny, *xii*
Brown, Patrick RJ, 205
Brown, Rosemary, 35, 50 n.32
Butler, Orvy, 194

Cage, John, 115
Capitani, Luciano, 30
Captain Midnight (MacDougall), 167
Carlos, Wendy, 117
Carpenter, John, 221
Chladni, Ernst, 196
Clapton, Eric, 7
Cohen, Jeffrey, 13
Colvin, Barrie, 14-15
Compton, Arthur Holly, 86
Connor, Steven, 12-14, 29, 43
Cook, Florence, 190
Cotard, Jules, 198
Criss, Peter, 114
Crocker, Mrs. H.H., 36
Crookes, William, 34, 190
Crosby, Bing, 191-82
Cross, Lowell, 115

Davis, Frank, 120
Davis, Rick, *xii*
de Lafforest, Roger, *xiii*
Derrida, Jacques, 171, 176 n.29
Deutsch, Herbert, 217-20, 222, 223-24
Dickinson, Bob, 171, 173
Dickinson, Wiliam Austin, 72 n.22
Draper, Henry, 234-35
Dryer, Ivan, 116-17
Duckwitz, William, 87

Eberle, Edward W., 66
Edison, Thomas, 4, 26-7, 28, 58, 79-80, 99, 197, 233-37
Eliason, Robert, 223
Elliman, Paul, 127-28

Farr, Mel, 262
Farrar, John, 93
Fawkes, Guy, 164
Feaster, Patrick, 9 n.3, 69
Fisher, Mark, 171
Flournoy, Théodore, 63
Floyd, Charles "Pretty Boy", 88
Flynn, William, 99
Fournel, Nicolas, 69
Fox, Kate and Margaret, 47
Fox, Marshall, 235
Franz Joseph I, Emperor, 179

Frewer, Matt, 163
Friedman, William F., 67

Gagnon, Richard, 128, 130, 133
Galbraith, Robert M., 234-36
Garmire, Elsa, 116-17, 123 n.23
Garson, Mort, 220-21
Gassett, William, 87
Gaye, Marvin, 262
Gee, Grant, 171
Gell, Alfred, 144
Gibb, Russ, 6-7, 62
Gifford, Denis, *xiii*
Glanvill, Joseph, 32
Goethe, Johann Wolfgang von, 31, 39, 50 n.44, 187
Goodman, Steve, 194, 198, 206
Gordon, James P., 106, 121 n.2
Grant, Ulysses S., 25
Graves, Michael, 261
Grierson, Francis, (pseudonym), 25ff, 48 n.2, *see also* Shepard, Jesse
Guins, Raiford, 12

Hammond, John Hays, 262
Hardy, Phil, *xii*
Harriman, W. Averell, 94
Harryhausen, Ray, *xii*

Hawking, Stephen, 133
Hawkins, Gay, 12
Haynie, Thomas, 164
Hernandez, Jackie, 228
Heys, Toby, 194, 198, 206
High, Sam, 38
High, William and John, 38
Hilderbrand, Lucas, 167
Hill, Walter, 132
Hitler, Adolf, 180
Home, Daniel Douglas, 33-34

Jackson, Shirley, 264
Jagger, Mick, 221
Javan, Ali, 105, 108, 110, 122 n.10, 122 n.11
Jefferies, Carson, 115
Jenkins, Charles Francis, 57-70
Jourafsky, General, 36
Jürgenson, Friedrich, 10-11, 187

Kaprow, Allan, 238 n.9
Kelly, Mervin, 121
Kennedy, Patrick, 143
"King, Katie," (spirit) 194
Kinsel, Tracey, 116
Kittler, Friedrich, 4, 13, 14, 27, 49 n.13, 63
Klein, Rick, 166, 175 n.13
Kneale, Nigel, 10, 133
Kulik, Leonid, 225

Latour, Bruno, 7, 65
Lauder, John, 136
Lear, John, 95
Leopold (spirit), 63
Lethbridge, Thomas (T.C.), 9-10
Liszt, Franz, 34, 35, 51
Littler, Richard, 170
Lockyer, Norman, 235
Lowell, Percival, 65-66

MacDougall, John R, 167-8
Maiman, Theodore, 108, 119, 122 n.3
Marie Antoinette, 32
Marin, Carol, 162-63
McCartney, Paul, 6
McDonald, John, 226
McLuhan, Marshall, 113
Meniere, Prosper, 156
Meyers, Professor, 235
Mompesson, John, 32
Moog, Robert, 217-19
Morin, James, 252, 254
Morton, Henry, 235
Morton, Hypolite, 194, 207 n.1
Morton, Jack, 120-21, 128
Mozart, Wolfgang Amadeus, 34, 35, 39
Mullin, Jack, 180-82, 183 n.6
Mulroney, Brian, 157
Nicholson, Alexander, 204

O'Brien, Kathleen, 132
Ogdy (thunder god), 201

Paglen, Trevor 207, 205 n.31,
Parikka, Jussi, 14
Parker, Magdalena, 204-206
Pfleumer, Fritz, 180, 201
Piccard, August, 88
Piccard, Jean and Jeannette, 86-88
Pierce, John, 113
Poulsen, Valdemar, 179-80, 182, 197
Prokhorov, Alexander, 108

Raimi, Sam, 151-52
Rammellzee, 111
Rathenau, Walther, 30
Raudive, Konstantin, *xiii*, 11, 186, 190
Reed, Thomas B., 28
Reynolds, Simon, 171
Roan, Dan, 158
Rockwell, James, 116
Roddenberry, Gene, 88
Rogo, Scott, 30

Sachs, William, *xi*
Saliger, A.B., 27-28
Scandura, Jani, 12
Schaeffer, Pierre, 31
Schawlow, Arthur, 106-8, 122 n.4

Schüller, Eduard, 194, 196, 201, 203
Schwartz, R.N., 119
Sconce, Jeffrey, 49 n.13, 169
Seley, Jason, 217
Serios, Ted, *xi-xii*, *xiii*
Shepard, Jesse, 25-28, 34, 35-47, 48 n.2, 50 n.44
Sherman, Donald, 131
Slepian, Walter, 194, 197, 202
Smith, Hélène, 63
Squier, George Owen, 80-83
Stanley, Paul, 114
Stone, Allucquère Rosanne, 133
Strauss, Johann, 117

Tandy, Vic, 199, 208 n.11
Taylor, Daniel, 248-49, 251, 254
Taylor, Frederick Winslow, 82
Tedworth Drummer, 31-32
Thalberg, Sigismond, 41
Theremin, Léon, 94, 217
Thompson, Emily, 68
Thompson, Stith, 19 n.1
Tilley, Christopher, 13
Todd, David, 64, 67, 69, 70, 72 n.22
Todd, Mabel Loomis, 72 n.22
Tompkins, Dave, ix-xv, 127, 257-65
Townes, Charles Hard, 106-7, 108, 119, 121 n.22, 122 n.4,

123 n.23, 124 n.30
Tudor, David,115
Turing, Alan, 202
Turner, Frederick Jackson, 63
Tyrrell, George N.M., 186, 189

Usami, Keiji, 115-16

VanDerBeek, Stan, 123 n.26
Vent, Bonnie, 45
Vrillon, 159-60, 170

Watson, Craig, 235
Watson, William, 4
Wesserman, H.M., 171-72, 187-88,
Whitman, Robert,115
Wilhelm II, Kaiser, 98
Willin, Melvyn, 30
Wilson, Woodrow, 98
Wisniewski, Prince Adam, 40
Wojcik, Daniel, 260
Wozencroft, Jon, 171
Wren, Christopher, 32

Zebb, Patricia, 242
Zeiger, H.J., 106, 121 n.2
Zworykin, Vladimir, 93-95

ACKNOWLEDGEMENTS

With special thanks to those who have read and edited things, provided encouragement, opportunities, and new connections—each contributing in wildly different ways: Bernie Brooks, Dave Tompkins, Marc Greuther, and Steve Aldana. With thanks to faculty who have had an especially great impact on my creative and critical processes: Stanley Rosenthal (RIP) and Evan Larson at Wayne State University; Daniel Wojcik and John Fenn at University of Oregon; Norman Bryson, Mariana Wardwell, Lesley Sterne, and Brian Cross at UC San Diego. Thanks to my colleagues at The Henry Ford, who think through things with me every day, and whose support along the path to becoming a curator has been paramount. Additional thanks to Herbert Deutsch, Paul Elliman, Grant Wythoff, and Christopher Laursen. Shout-out to Mark Fisher in the hauntological afterlife. Thanks to the creative teams at Moogfest, Unsound, and Pop Kultur, three festivals that afforded me the opportunity to present this sonic research in its early phases. Of course, to my publishers at Strange Attractor Press and MIT, who saw value in this book from the start: Mark Pilkington, Jamie Sutcliffe and Matthew Browne. And definitely not least, though last in this list, thanks to friend, patient editor, and champion of basset hounds, James Hughes.

STRANGE ATTRACTOR PRESS 2019